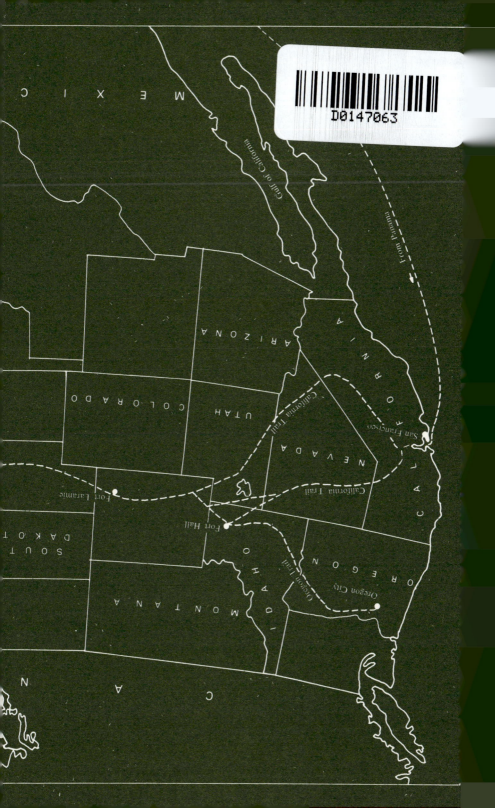

203, 213, 251, 254, 255, 260, 264-5
Shull, Jesse, 29
Sierra Nevada, 10, 50, 51, 54, 72, 73, 77, 103, 106, 111, 112, 146, 152, 155, 184, 185, 194 ,232
Silicosis, 56, 75, 124, 154, 162, 168, 173, 174, 204, 207, 227, 237, 245, 246, 263
Slavery, 45, 80
South Africa, 182, 236, 237, 245
Southampton, 116, 237, 263
St. Lawrence River, 36, 45
St. Louis, 28, 29, 45, 60, 233, 248
Strikes in industry, 70, 72, 94, 129, 131, 132, 161, 163-5, 166, 167, 180, 188, 219, 228, 229, 237, 244, 268-9
Strike, John, 144 6
Superintendents, Cornish Mine, 64, 74, 76, 77, 81, 87, 88, 92, 93, 103, 107, 115, 118, 119, 123, 124, 126-7, 132, 137, 138, 140, 141, 142-3, 145, 149, 153, 161, 167, 174, 178, 182, 186, 188, 194, 203, 219, 225, 226, 245, 254, 260, 262, 263, 264
Sutro, Adolph, 189
Sutter, John Augustus, 50, 51
Sutter's Fort, 50, 51
Swansea, 153, 160, 254
Swindles, 111, 159, 163, 193, 258

Talkalai, Chief, 265, 268
Taylor, Bayard, 61, 161
Techniques of Cornish mining, 16, 22, 23, 47, 48, 52, 54, 55, 56, 67, 68, 71, 72, 77, 85, 90, 91, 94, 108, 117, 118, 119, 128, 160-1, 264, 268
Teller Opera House, The, 25, 148, 157, 158
Trails to America, 21, 22, 34, 36, 37, 43-46, 56, 58, 59, 60-1, 67, 72, 73, 75, 76, 77, 85, 86, 87, 104, 107, 109, 116, 118, 126, 130, 152, 155, 161, 173-4,

175, 181, 204, 210-15, 216, 217, 229, 231, 233, 234, 235, 237, 245, 248, 263, 265-7
Transvaal, The, 144
Universities and Colleges, 53, 71, 77, 93, 113, 115, 123, 143, 144, 145, 149, 150, 215 216, 217, 234, 266, 267
Utah, 104, 217, 229, 232, 233-241, 252

Venezuela, 102
Victoria (British Columbia), 15
Virginia, 14, 152

Wages, in America, 60, 61, 62, 63, 67, 70, 72, 89, 90, 91, 92, 94, 95, 102, 116, 120, 127, 128, 129, 131, 139, 149, 163, 165, 166, 167, 174, 182, 188, 189, 202, 203, 219, 232, 237, 255, 263, 264
Wages, in Cornwall, 17, 18, 237
Warren and Johnson (publishers), 14
Washington, 104, 215, 222, 245
Wells Fargo, 51, 223
White, Peter, 138
Wisconsin, 28 50, 59, 61, 62, 63, 69, 77, 110, 118, 120, 142, 174, 209, 210, 216, 227, 229, 248, 254, 256, 266
Work, Conditions of in America, 74, 77, 78, 80, 91, 94, 95, 96, 102, 108, 118, 119-20, 120-6, 128, 129, 138, 139, 148, 151-2, 155, 160, 166, 174, 176, 181, 183-207, 219, 237, 247, 249, 250, 251, 252, 255
Work, Conditions of in Cornwall, 10, 15-24, 104, 116, 128, 130, 143, 144, 145, 146, 148, 153, 181, 230-1, 237 246, 263, 265
Wrestling, 71, 73, 101, 139, 141, 178, 190
Wyoming, 222
Young, Brigham, 234

THE
CORNISH MINER
IN AMERICA

the contribution to
the mining history of the United States by
emigrant Cornish miners—the men called Cousin Jacks

ARTHUR CECIL TODD

B.A., PH.D.

D. BRADFORD BARTON LTD · TRURO
THE ARTHUR H. CLARK CO · GLENDALE

331.116822
T633c
1967

First published 1967 by
D. BRADFORD BARTON LTD
18 Frances Street, Truro, Cornwall
and
THE ARTHUR H. CLARK COMPANY
1264 South Central Ave., Glendale, California

Copyright © Arthur Cecil Todd 1967
All rights reserved
LIBRARY OF CONGRESS CATALOG CARD NO. 67-26342

Printed in Great Britain by
H. E. WARNE LTD. ST. AUSTELL CORNWALL

CONTENTS

PREFACE 9

I CORNWALL AND ITS MINERS 13

II WISCONSIN AND ITS CORNISH "BADGERS" .. 28

III CALIFORNIAN GOLD, MERCURY AND SILVER .. 50

GRASS VALLEY, NEVADA CITY AND THE NORTHERN
MINES 54

THE NEW ALMADEN QUICKSILVER MINE .. 79

THE SOUTHERN MINES, THE MOJAVE DESERT, DEATH
VALLEY AND YOSEMITE 103

IV THE COPPER AND IRON WILDERNESS OF UPPER
MICHIGAN 114

V COLORADO AND "THE RICHEST SQUARE MILE ON
EARTH" 151

VI HUNTING SILVER TO THE DEATH IN NEVADA .. 183

VII BEYOND THE ROCKIES TO THE PACIFIC NORTH-WEST 208

VIII BACK TO COPPER: UTAH, MONTANA AND ARIZONA 233

INDEX 271

ILLUSTRATIONS

Section of a lead mine in Wisconsin in the 1840's 65

Emigration broadsheet issued in Padstow in 1844 66

The Miners' Union Hall at Bodie, California 99

The Cornish town of Nevada City, about 1890 99

The young Richard Jose of Lanner singing carols in Virginia City .. 100

A shelter built by Cornish miners at Calico 100

The remains of a Cornish pump in the Cora Blanca Mine, New
    Almaden 133

"Crowst-time" for Cornish miners at New Almaden in the 1980's .. 133

Pay day for Cornish and Mexican miners at New Almaden .. 134

Section of the Cliff Mine, Lake Superior, about 1850 135

Surface workings of the Cliff Mine, about 1850 135

A mining ballad commemorating the Central Mine disaster .. 136

The famous Chapin Mine at Iron Mountain, Michigan 169

The Chapin Mine's enormous Cornish pump 169

A drilling contest at Eldorado, Colorado 170

Timbering in a mine at Central City, Colorado 170

St. Just-in-Penwith, looking towards the Atlantic 171

Wardner, Idaho, in the days when it was a mining settlement .. 171

Rocky Bar, Idaho, as depicted in 1880 172

Six-Mile Canyon and Virginia City, Nevada, about 1900 205

Silver City, a Cornish town in Nevada, as it looked in 1890 .. 206

Part of the industrial complex of Virginia City 206

The chapel at Alpine, Utah, built in 1863 by John Rowe Moyle .. 239

Pascoe's livery stable at Globe, Arizona 239

The very first bank in Miami, Arizona 240

The Old Dominion smelter at Globe, familiar to thousands of
    Cousin Jacks 240

MAPS

Cornish settlements in Northern Michigan and Southern
    Wisconsin   33

Cornish settlements in California and Nevada   57

Main Cornish settlements in Colorado and South Dakota   ..   159

Cornish settlements in Southern California, Utah, and Arizona   ..   187

Cornish settlements in Montana and Idaho   220

To Cousin Jacks everywhere
and especially to W. J. Ellery
who was so enthusiastic about this book
but died in 1965 before it was completed

PREFACE

THE idea of writing this book came to me many years ago, and many hands and voices have been involved in its preparation. When I came to Cornwall just after the war I was astonished and excited by two phenomena: the fallen monuments to its great industrial past set against a landscape and seascape unsurpassed anywhere in England; and the fact that almost every Cornish family had relatives in America. Cornish students in my university extra-mural classes would reveal quite casually some surprising information; one, who was a bus conductor, had worked in Detroit; another, a housewife, had been to school in Michigan; a third apologised for being absent for a term as she was returning to Idaho to see the silver camp where she was born; and a fourth claimed that he was the cousin of Deadwood Dick.

To a Celt like myself from the other side of the border it was bewildering to live and work in a community that was half Cornish and half American in its outlook. One could talk to an old miner on a harbour wall and discover that he had never been to London but could describe quite clearly the streets and taverns of San Francisco. Some of the Cornish in the extreme west even spoke like Americans. Yet very little seemed to have been written about this extraordinary situation. Books there were in plenty about the copper and tin mines, the technology of mining, the engine houses with their massive beam engines for working the pumps, Cornish inventors like Trevithick and Hornblower, and the miners themselves and the frightful conditions under which they worked. But the central tragedy of their lives, the closing of the mines and the hard decision that often had to be made between entering the dreaded workhouse or emigrating to America, appeared to have been largely ignored by historians. In a way this is understandable for the folks who were worst hit by the depressions of the last century were ordinary men and women with little or no formal education, unlikely to record their trials and tribulations in letters and journals. Yet it seemed to me that there could be no complete understanding of the emigration movement that concerned the passage of more than 20,000,000 souls from Europe to America until we knew something of the personal lives of the emigrants themselves.

I have tried to fill this gap by discovering what happened to the

9

Cornish miner and his family *after* they had arrived in America. Some critics may say that such accounts are not eventful and dramatic enough to warrant this special treatment and that these people are only the "little men" of history, a view I am sure that the miner would be the first to admit. Yet the fact still remains that it was they who daily had to face strains and stresses of life that the more fortunate escaped: inadequate houses, poor food, disease and the perils of their occupation, to which were now added the problems of conquering an unfamiliar and often hostile environment. That they adapted themselves successfully demonstrates that they possessed virtues which amounted to the heroic. Poverty of spirit could never have ensured a safe crossing of the Sierra Nevada under a burning sun or surviving a long winter in a howling wilderness by Lake Superior. Their penetration and settlement of the hard-rock regions that stretch across the American continent in an immense sweeping curve from Michigan in the north to Arizona in the south-west is an epic story. They left behind wives and children, born and unborn, in a cold, damp and disease-ridden Cornwall; they travelled steerage on borrowed money; they picked open mountains; they trod trails across deserts; they made millionaires out of the sweat of their hands and backs; they brought law and order to rough mining camps by building their churches and schools; and they pioneered the mining industry of America. No mine was ever without its Cornish captains and every mining company was keen to employ them, often to the disgust of their neighbouring Celts, the Irish. Nor was any Cornishman ever without a job, and, since there was always some "cousin" in Cornwall waiting for work, the American miner and those of other nationalities called him Cousin Jack.

The telling of this tale is largely their work, helped by a sprinkling of farmers from the Lizard. Since such workers rarely talk about themselves or get themselves into print, information at first had to be acquired at long range by personal correspondence with their descendants. I owe an immense debt of gratitude to hundreds of these Cornish-Americans who, pestered time and time again by my questioning, sat down to the labour of writing, hunted for lost relatives and entrusted me with precious family portraits and documents. Then, when the opportunity came for me to visit America, they entertained my wife and myself; met us off trains at dawn, bedtime or in the middle of the night; ferried us in their cars to inaccessible ghost towns in mountains, deserts and forests where, paradoxically, the only signs of life were the inscriptions on weathered Cornish tombstones in countless cemeteries. One of the lasting pleasures of this kind of research has been the many friendships formed; but inevitably, with the passing of the years, some have not lived to see the work completed, like Mabel Thomas of Oakland, a very dear and helpful friend, whose father came from Porthtowan.

Many of these Cornish-Americans are mentioned in the main body of the text, but if there are any omissions, I hope I may be forgiven. There are others I wish to thank for special guidance and advice: Mr. and Mrs. Laurence Bulmore of San José, California; Mr. Clyde Arbuckle, the City Historian of San José, and his wife; Mrs. W. A. Klaus of Philadelphia; Mr. and Mrs. N. Carter of Boise, Idaho; the late Mr. W. J. Ellery and his wife of Monrovia, California; Dr. and Mrs. Alton Moyle of Madison, Wisconsin; Mr. and Mrs. Herbert Hamlin of Sonora, California; Raymond Carlson, the editor of *Arizona Highways;* Mr. and Mrs. Olen Switzer of Oregon City; Mr. and Mrs. Walter Dry of Manzanita, Oregon; the late Matt Kelly of Butte, Montana; Wilfred Nevue of Champion, Michigan; the California Pioneers of Santa Clara County; Mr. and Mrs. Thomas Foxwell of Elgin, Iowa and Norma Hage of Los Gatos, California.

The University of Exeter gave me a year's leave of absence, the Fulbright Commission a Travelling Fellowship and the Leverhulme Trust a generous research grant: to all of these bodies I am especially grateful. The University of California at Berkeley, where most of the research was carried out, was lavish in its hospitality and invited me to teach in two of its departments; and Dr. Paul Sheats, Professor Henry Nash Smith and Professor Mark Schorer made my stay on their campus for two semesters a most enjoyable experience. American librarians have been at all times most helpful; the Director and staff of the Bancroft Library, Berkeley; Mr. H. J. Swinney of the Idaho Historical Society; Mrs. E. Stones of Albuquerque; Margaret Gleason of the Wisconsin State Historical Society; Miss Helena Symons of Lawrence, Kansas; the late Mrs. Carroll Paul of the Marquette County Historical Society, Michigan; Mrs. Ruth Newport of the Tuolumne County Historical Society; Mr. Robert Neal of the Mineral Point Historical Society, Wisconsin; and Mrs. Alys Freeze of the Denver Public Library. Mrs. Clara Woody of Globe kindly read the section on Arizona and made several valuable suggestions. Mr. Jess Hayes, with whom we toured the Apache Reservation of San Carlos, placed at our disposal the records in the Court House of Globe. Mr. Ben Coil of the Miami Copper Company cleared up many technical points about copper mining today; and Miss M. E. Redman has read the entire manuscript. I owe a special debt of gratitude to my publisher, Mr. D. B. Barton, for his expert technical advice about the final shape of the book; and to Mrs. R. M. Barton, for her consummate skill in drawing the maps, which are a model of draughtsmanship.

Lastly I would like to thank my wife for her constant encouragement, for her many ideas and suggestions and for her untiring pruning of the several drafts of the text. She has been deeply involved in this study at all stages in its development, enduring the fatigues, and enjoying the fun, of travelling 24,000 miles with me to inspect scores of sites, and interview hundreds of people and photograph them. The book is as much hers as mine. We both realise the holes in the narrative that need filling. What has been attempted is no more than a cross-cut along a subject that has no end while our Cornish friends have to leave their county in search of work. We did not explore Montana, the Dakotas, Washington or Minnesota, nor the copper regions of the Appalachians. And there are many more Cousin Jacks in America whom we ought to have visited. But "ars longa, vita brevis" is our excuse.

A. C. TODD

Penzance, Cornwall
July 1966

I

CORNWALL AND ITS MINERS

CORNWALL opens a window on the world in a way that no other county of England does. Geographically it points to the New World, while for the touring American its off-shore rocks and lighthouses are the first glimpses he has of England. Here painters find a paradise of colour where a brilliant light is reflected from sea and cliff to the sky; sculptors see a panorama of rolling uplands with weathered granite cairns and headland profiles which have surely influenced their techniques. But Cornwall is also a realm of legend and ancient history: of the mythical kingdom of Arthur, the sad tale of Tristram and Iseult, and the lost cities of Lyonesse; of the Iron Age warriors and their hill-top forts; and of the old Bolerium of the Romans who sailed along its misty coasts from Spain. It is both a princely province of medieval castles and a Royal Duchy: the eldest son of the ruling monarch of England is Duke of Cornwall from the moment he draws his very first breath. It is also, and has been since the time when men first roamed the sea, the grave-yard of mariners who have ventured too close to its treacherous shores. It is the land's end and the ocean's beginning, a peninsula of granite and slate, of green pastures and golden gorse, that has encouraged its dark stocky Iberian race to take the world in its stride. Yet wonderland that Cornwall is, instilling a magic in every traveller bound for his "south-west discoverie", it is also a wounded land of red gashes, blackened sores and seeping cuts that have drained its life away. Here, within an area of less than 2,000 square miles that could be tucked away comfortably into northern California, the secrets of mining near the earth's core were locked for centuries along with the wisdom of the countless saints who built their rough baptistries by the banks of its streams. Then suddenly in a relatively brief moment of time its people were scattered throughout the world and the land left wasted and blighted.

The Cornish of course have not been the only people to mourn this kind of disaster. It has been estimated that in the century between 1815 and 1914 no less than 35,000,000 left Europe to better themselves in North America, Australia, New Zealand, South Africa and elsewhere. This phenomenal army of souls on the march to a promised land constituted a movement so widespread in its origins, so diverse in its results and so complex in its nature that no final conclusions can be drawn about it until the special influences at work in every country have been considered. Moreover, the emigrants themselves were generally humble and obscure men and women who, cast in an heroic mould as they must well have been, since they faced up to the terrors of a sea crossing and made the adjustment to an unfamiliar and often hostile environment, had no pretensions to be remembered outside and beyond the span of their own lives. For the most part the rank and file who contributed to the making of modern America are relatively unknown except in the most general terms of being either courageous, enterprising and self-reliant or rapacious, arrogant and selfish. Thus all too often emigration has been studied from the lives of the more exceptional who came, saw and conquered.

Cornwall has its roll of famous men, and their praises have already been sung on both sides of the Atlantic. Charles Morton, the son of a rector of St. Ive, in 1685 was invited to become the Vice-President of Harvard.[1] Samuel Penhallow of St. Mabyn first worked as a missionary among the American Indians and then became a merchant, judge and historian[2]. Jonathan Hornblower of Penryn in 1753 built the first steam engine in New Jersey, erected it himself, vowed he would never cross the Atlantic again and in 1785 represented New Jersey in the Congress of the Confederation[3]. John Grigg, born in 1792 of farming parents in the parish of St. Stephen, after a varied life in Ohio and Virginia, acquired the publishing firm of Warren and Johnson and made it the most prosperous in the United States. In 1850 he sold it for a handsome profit and engaged in private banking and western land speculation; but he never forgot his relatives in Cornwall, either financing their emigration to Sangamon County, Illinois, or remitting cheques to those left behind in Trewint, Launceston, St. Austell, Redruth and Charlestown to the extent of 250,000 dollars[4]. The botanist William Lugg of Egloshayle, after searching the jungles and mountains

---

1   See F. L. Harris, 'Charles Morton—Minister, Academy Master and Emigrant (1627-1698)'; and F. A. Turk, 'Charles Morton; His Place in the Historical Development of British Science in the Seventeenth Century', *Journal of the Royal Institution of Cornwall* (New Series, Vol. IV, part 3, 1963) pp 326-363.
2   See *Dictionary of American Biography* and *Collections of the Massachusetts Historical Society* (2nd Series, Vol. 1) p. 161.
3   See William Nelson, 'Josiah Hornblower and the First Steam-Engine in America', *Proceedings of the New Jersey Historical Society* (Vol. VII, No. 4, 1883) pp. 175-247.
4   There is no biography of Grigg, but some information may be found in Stephen Winslow, *Biographies of Successful Philadelphia Merchants* (Philadelphia: 1864). I am grateful to Mr. James Taylor Dunn, the Librarian of the Minnesota Historical Society, for drawing my attention to Grigg, who is one of his ancestors.

of South America for orchids and conifer seeds, in 1849 was threading his way among the gold hunters in San Francisco bound, not for the Mother Lode and its camps, but for the forests of redwood and Douglas fir; back and forth he travelled between London and San Francisco, dying within sight of his Pacific and completely indifferent to the movement of the miners. And then there were the Penroses of Helston, linked to their shipbuilding brothers of Bristol, who settled in Philadelphia and produced generations of engineers; Charles Penrose, who died in 1958, distinguished himself as President of the Maryland Academy of Sciences and Vice-President of the Newcomen Society of North America, of which he was one of the founders.

But, as George Eliot remarks in *Middlemarch:* "the growing good of the world is partly dependent on unhistoric acts; and that things are not so ill with you and me as they might have been, is half owing to the number who lived faithfully in hidden life and rest in unvisited tombs". Who visits now the grave of the Quaker saint of St. Columb Major, John Gilbert, who was imprisoned with George Fox in Launceston jail and in 1682 emigrated to Philadelphia? Who now remembers the genial Matthew Moyle and his assistant Nicky Trevathan who superintended the crews of Cornish, American and Negro miners in the little known Gold Hill mine in North Carolina?[1] Yet all over Cornwall buildings and landscape evoke memories of the American odyssey: Dakota Farm, Nevada Terrace, Michigan House; cottages named Monrovia and Shasta; the Mexico Inn near Penzance; Gold Hill fields over Carneel Downs near Helston; a section of Redruth known as "Cally" and a street called Chili. Though these are more numerous than Cornish place-names in America, yet one remembers Pierce and St. Just Hill in Arizona and Barkerville in British Columbia, immortalised by a Cornish mining seaman who in 1858 jumped ship at Victoria, headed for the Fraser River and made a fortune of 500,000 dollars in gold.

With these names are associated a series of skills and techniques from one of the oldest mining regions in the world. They were perfected by a system that at one time permitted a son to work beneath the watchful eye of his father when he was only seven years old; in this way the "mathematics of the mole" were transmitted from one generation to another. The miner was the hunter; and his mine provided all the excitements of the chase as the lodes twisted and writhed and then disappeared. But the mine was also his father and mother, for his whole

---

[1] See *The Salisbury Sunday Post*, 10 August 1952.

personal life was bound up with it as in no other industry. For some, mining was almost a religion; they really believed they were working for Providence and helping mankind; thus they referred to their earnings as "clean money". This claim to a special position in the hierarchy of the world's workers was perhaps justified since they were by far the most sophisticated exponents of the art of mineral extraction. Long ago they had solved all the practical problems of blasting, timbering, hoisting, flooding and ventilating. They had evolved a language that became a *lingua franca* in mining camps the world over: pits were shafts; shafts from one gallery to another were winzes; horizontal galleries excavated in mineral veins were levels; galleries in rock were cross-cuts; tunnels for drainage were adits; downward excavations were sinkings and upward excavations were risings. Ore that was deliberately left in the mine as an investment against future depressions was known as "the eyes of the mine"; and the process of extracting it was termed "picking out the eyes", a powerful poetic image.

By 1850 Cornwall was still producing much of the world's supply of copper and tin and its prospects seemed almost boundless. Because of the lack of properly conducted surveys it is difficult to say how many mines were working or the size of their labour force. Spackman's *Analysis of the Occupations of the People* of 1827 estimates that in copper mining there were 11,639 males and 2,098 females; in tin mining 5,706 males and 130 females; and that these together with a small number engaged in lead and iron extraction accounted for a total of just over 20,000. Within the next thirty years, however, this figure had more than doubled for, according to Williams' *Mining Directory* of 1862, 340 mines were then employing 50,000 Cornishmen and women. For the miner life was nasty, brutish and short, with endless physical toil and often marginal poverty, eight hours a day being spent underground except mercifully on Sundays. The sun shone on them only infrequently; their complexions were wan and sallow; their bodies were stunted with crawling on their knees along the narrow galleries and in the same cramped position hammering through the granite at the rate of no more than a foot a day by the light of a single candle. No less exhausting was the effort expended in reaching the rock face. In the days before the man engine, the only means of ascent and descent was by almost vertical ladders that were always wet and slimy and often had rungs broken or missing, a method of locomotion which used up as much as one fifth of their energy. Some mines were so deep that a man might have to climb continuously for an hour before he arrived "at grass", that is at the surface; and during this time it was more than likely that he would pass from temperatures of 80°F and high humidity to the icy air of the main shaft before he began his walk home, often in the teeth of one of the Atlantic gales for which Cornwall is notorious. After his long stint underground, confined within a narrow stope where the air was often fouled by the stench of human excreta, a miner's heart and lungs were never in good shape, and their gradual deterioration was a certain cause of accidents. Yet the loss of an arm or a leg might actually be welcomed since he then would be able to work at some surface job in the sunshine, whereas the alternative was a slow and cruel waiting for death sitting in front of an open window coughing and spitting his hours away, his lungs having been punctured by the granite's fine dust and further weakened by the damp Cornish air. Few constitutions could stand the strains of mining; more than half of the miners contracted silicosis; their working lives were finished before they were forty and few ever reached the age of sixty. The miner's cottage on the windswept moor might be his castle but it was defenceless against the rigours of his employment. It is true that he might own it, having built it with his own hands from the rough stone that lay around, or rent it for as little as £3 a year; that he had the right to cut turf which gave his family a cheap, though poor, source of heat; and that he had filled the wall cavities with a mixture of clay and dung that helped to retain the warmth. But in the only two rooms upstairs there were no ceilings; as the family lay in their beds they could see the stars through the gaps in the shingles and feel the night wind on their faces. Nor was the diet adequate and nourishing for the fight against disease: potatoes for padding and pilchards for the sustaining oil. Both could be plentiful and cheap, but if the harvest even partially failed starvation could result; blight could ruin the one and some whim of nature take away the other. The effect on children could be catastrophic; dirt, disease and the lack of sanitation took their toll of young lives; at St. Just-in-Penwith in the years 1840-49 a quarter of all males buried and a half of all females were boys and girls under the age of five[1].

Conditions were barely tolerable even when times were good; when they were bad the worn-out miner with heart disease dreaded the day when his only refuge would be the Workhouse. In periods of prosperity a tributer or contract worker might earn between £3 and £4 a month and an engineer as much as £8, but the average wage was more likely to be as low as £2, from which sixpence a week might be deducted for insurance against sickness, though this was not the general practice.

---

[1] John Rowe, *Cornwall in the Age of the Industrial Revolution* (Liverpool: 1953), p. 152.

It was only in well-managed mines like Fowey Consolidated or Gwennap Consolidated that a miner received thirty shillings a month sick pay together with medical attention for himself and his family. For the most part a family's survival depended on every member working at the mine, even the young children. Boys and girls were only seven when they entered the labour market. On the surface they separated the "deads", that is the rubbish, from the ore and earned three or four pennies a day for a ten-hour shift in the summer and a nine-hour shift in the winter with a rest period of no more than one hour which they usually spent in the blacksmith's shop when the weather was cold and wet. William Rodda, who was born in 1845 and died in New Jersey, remembered working underground with his father at Dolcoath at the age of seven retrieving wooden wedges that were used to shore up loose rock. There he would stumble with a heavy wheelbarrow or turn the borer that was being driven into the granite, his wages about thirteen shillings a month. His employer was invariably his father or an uncle, for the mine adventurers, that is the shareholders, were artful enough not to undertake the responsibility of placing such young people on their pay rolls. Later on, as his body grew in strength, the tributers would take him into partnership and then his earnings would vary with theirs; but once in the mine there was no escape.

Girls, on the other hand, never went underground though their work was just as exacting and always manual, few of them being seen without a stone mallet in their hands. The older women, many of them mothers, hammered the rude ore to pound it down to a smaller grade and then passed it on to the "bal maidens", adolescents of sixteen or so who "bucked" the ore with a flat iron hammer to reduce it to the size of half-inch marbles. Their wages, according to the returns of the Consolidated Mines in the parish of Gwennap, were graduated according to age; 12 to 14 years, twelve shillings a month; 14 to 17, fifteen shillings; and eighteen shillings for all over seventeen. Stooping over the piles of ore apparently affected the figure of the Cornish girl as one observer noticed: "The use of hammers in dressing ores tends, perhaps, to the production of some fulness of breast, but the sedentary position necessary gives little or no exercise to the lower limbs". But without their meagre earnings life for most families would have been unendurable, for all knew that mining was a lottery in which there were few prizes.

The Cornish industry suffered also from being grossly inept and wasteful in its management; too many small adventurers were draining too many small and uneconomic mines; and the custom of the annual share-out of profits, rather than declaring dividends, meant that they were hardly ever ploughed back for improvements. By the late 1850's high grade copper was showing signs of pinching out at a depth of 1,000 ft and, as if this was not serious enough, was facing severe competition from the vast and cheaply worked deposits that had been discovered on Lake Superior and in Chile, resulting in a lowering of world prices. How devastating Michigan copper was for Cornwall can be appreciated from its annual production figures: in 1845 a mere 24,880 lbs; by 1855 a sharp jump to 5,809,334 lbs; by 1865 14,358,592 lbs; and by 1875 a staggering total of 36,039,490 lbs. Against this tidal wave the Cornish adventurers were completely helpless. The American Civil War, it is true, granted them a short respite, but by 1865 the writing was clearly on the wall, even though some adventurers believed they had bought time by changing from copper to tin; but they too were hit hard again when richer deposits were discovered in Bolivia and Malaya.

The only solution to the widespread unemployment and human misery that resulted from the closure of mine after mine was emigration. How many embarked on this path of no return it is almost impossible to calculate with any accuracy without a searching survey of the county parish by parish. But the Registrar General, in his report on the census of 1881, commented that in the previous ten years population had decreased by 8.9% and that it was possible that the mining population had shrunk by as much as 24%; L. L. Price of the Royal Statistical Society estimated in 1888 that one third of all the miners had left Cornwall for good; and Leonard Courtney, in a lecture to the Royal Institution of Cornwall at Truro in June 1897, reckoned that by the end of the century 230,000 would have departed. It seems reasonable to suppose that Cornwall lost at least one third of its population, which at the 1851 census stood at 341,269, and that the majority were the young men and women who took with them the best years of their lives.

Tragic though the separation was in almost every family after 1865, their future was never as hopeless as it seemed for those who went from continental Europe and Ireland; for the Cornish carried in their hands skills that were in unceasing demand on the other side of the Atlantic and so they were hardly ever to be without work. Moreover, in every Cornish household for many years North America was an even more familiar name than England. For instance, in the 1830's, faced with almost unprecedented destitution and the soaring costs of relief works, the Poor Law Commissioners had worked out a scheme of assisted passages to Canada for families who cared to apply. But even before that mining superintendents from Redruth had been trickling away with their assistants to manage the rich silver mountain masses in Mexico; and some of them were the first real miners in California before the gold rushes. Throughout the literary and scientific societies and the mechanics' institutes of Cornwall circulated such journals as *Chamber's Information for the People* and *The Emigrant and Colonial Gazette*, while newspapers like *The West Briton* regularly published emigrants' letters and gave reports about conditions in the mining camps. More significant, however, were visiting emigrants who came back to collect their families, the monthly remittances [1] and the pre-paid passage money that bought tickets in the offices of travel agents all over Cornwall. For instance, over the counter in the general stores at St. Just-in-Penwith it was possible to book right through to Houghton in Michigan; and on the occasion of a local feast day it was not unusual to see leaflets being handed out extolling the attractions of life in Arizona.

But by far the most telling event that was to speed up the whole process of emigration was the depression of the 1840's. The census of 1841, the year when parishes were first asked to furnish information about emigration, revealed that from January to June 795 people, or 2.3 per 1,000, had departed from Cornwall. They were mostly miners from Camborne and Redruth intent on reaching the lead regions of south-west Wisconsin, but with them were many educated and moderately wealthy farmers from the Lizard, bound for the rich prairies by Lake Michigan. Asking for no assistance or direction, they simply auctioned their farms and stock, arranged with friends in Wisconsin to buy them land, and plotted their own routes. Many were the reasons that drove them away: high rents, scarcity of labour, the unrewarding returns from the cultivation of marginal land, and increases in the poor rate and general taxation. Perhaps most important of all, they were tired of England and its class distinctions and hatreds; some openly talked of Cornwall as a "land of bondage" and only desired to taste the liberty that, from their own reading, they presumed existed in America. Politically and socially they were already to some extent Americans even before their ships left Falmouth.

The miners were never so fortunate as these yeoman farmers for they had nothing to sell but their labour and skills to pay for a passage. Cash would be raised by relatives to send out the young and strong or borrowed from tradesmen on the security of their cob cottages. It was

[1] *The West Briton* of 12 August 1869 reckoned that Camborne men in California and Nevada were remitting home to their families between £15,000 and £19,000 a year.

the realisation that debts had to be paid and that their families depended on them which sustained them rather than any preconceived ideas about liberty, utopia and the perfectibility of man. Nevertheless their experience stood at a high premium in the American mining market; and they had few doubts about rustling a job, especially if the mine captain happened to be, as he invariably was, a Cornishman. At a pinch there were other aptitudes a miner could offer for he was usually an excellent carpenter and mason as well. He was thus never likely to be found stranded in the boom towns of the eastern states of America or disconsolately wandering from place to place until he discovered the industrial niche that suited him best. When he set out for the West, it was towards a camp that in his mind and imagination he knew well already; and if he returned to the East, it was not as a failure, but to pick up the family that had been left behind. Pioneering broke many a man's spirit but rarely that of a Cornishman since he was called upon to pioneer in ground that was already familiar to him in Cornwall, hundreds of feet below daylight among the veins and lodes of a new mineral-rich mountain. Yet the hazards of his occupation were appalling, death in its most violent forms common, and the fatherless family a fearful probability.

The great exodus of the 1880's from towns and villages along the spine of Cornwall is first discerned as the sound of the distant rumbling of wagons in the 1840's down the rough roads and tracks from Camborne and Mullion to the nearest seaport of Falmouth, from where the mail-packets sailed weekly to New York. The destination of these first of the many was Wisconsin, its south-western edge the goal of the miners and the strand of Lake Michigan the destination of the farmers, some of whom were later to venture along the Oregon Trail to Willamette and beyond. The miners had no choice but to travel steerage, but this perhaps was not much worse than the steerage conditions of the Cornish cottage; the women cooked either in the stinking cabooses or on the open deck. There was no privacy, medical attention was nonexistent, and the captain was only obliged to provide water which was often not even fresh when rolled on board. In those early days they embarked on whatever ship happened to be lying in harbour and had to make their own arrangements with the captain who was often forced to crowd passengers beyond the safety limit because of the fierce competition between shipping lines and the cut prices that followed. In 1816 the fare from Liverpool to New York was about £12; in 1843 it had shrunk to £6 on packets and £4 on trading vessels; and in 1846 it had dropped to £3.

The routes to Wisconsin were many and almost equally dangerous, but the one by way of Quebec, although the cheapest, was the most prone to disaster because of fog, high winds and ice. Some attempted the long river route from New Orleans up the cholera-choked Mississippi; some trekked overland from New York and then followed the Ohio until it joined the Mississippi; others preferred to travel up the Hudson and then continue by barge through the Erie Canal to Buffalo, where they transferred to a steamer for what might be a pleasant voyage of several days across the Lakes if the weather was kind. Then, when California opened wide its golden gates, these Wisconsin pioneers were some of the first to appear in the golden canyons of the Mother Lode. If they could afford it, they rode the lead wagons to Galena in Illinois, boarded a river boat to New Orleans and fought for another ship to Panama. There they braved jungle and swamp, sometimes on the back of a mule but more often on foot; scrambled among the crowds of forty-niners for any vessel that would take them the thousand miles by sea to Yerba Buena, the beautiful City of the Spaniards later re-named San Francisco; and then trudged the final two hundred miles across the great plain of the Sacramento River to the diggings. The less affluent faced the rigours of the overland trail in their prairie schooners, daring mountains and high desert where the sky was forever hostile. And after 1865, as the tide of emigration mounted and new mines were opened, so the trails of these Cornish miners crossed and re-crossed through Michigan, Colorado, Nevada, Idaho, Utah and Arizona as they trod their pilgrims' way to the ore. Emigrants of other nationalities called them Cousin Jacks.

In later years when the Cornish had established themselves and displayed their real worth on the mining frontier, Cousin Jack was a term of endearment; but in the early days of struggling for survival it denoted envy, jealousy and even hatred for it seemed that every position in the mine was reserved for yet another "cousin" from Cornwall. This was inevitable for the Cornish were regarded, perhaps to their surprise, as a veritable industrial élite, an aristocracy of the pick and gad. Their mining lore and their mining terms, their kibbles, skips and whims, their long thin wheelbarrows designed for operating in the most confined spaces, the stope that was called "Cornish" because it was so narrow that only a Cornishman could work it, and their powerful beam engine that ponderously stroked its path up and down the shaft's darkness, everywhere proclaimed that here was a race of men to match the mountains of red, yellow, white and grey metals. In Cornwall they had once been the privileged possessors of their own Stannary Parliament with the right of exemption from certain forms of taxation; and in America they were equally privileged as company men, providing the superintendents, captains and drillers, and following in the wake of the placer and hydraulic miners. Technically they were the world's best hard-rock miners.

On the organization side of the industry they had also contrived to work out a system that took into account certain incentives to raise output. Its form derived from the very nature of shaft sinking and the construction of the galleries running away from the main shaft; these divided the lode into regular compartments about 300 ft in length and 60 ft in height, which were again sub-divided by small perpendicular shafts into "pitches", about 60 ft high and 30 ft long. The method of working these pitches was to put them up for competition, the miners contracting for a certain percentage to break the ore, bring it to the surface and, in some cases, dress it for the market; the ores were sold monthly and the miner received his "tribute" or percentage for which he had agreed to work. Sometimes a lode would pay well and sometimes it would vanish without trace; hence tributing was a business that required keen judgment and persistent application on the part of the miner who had to make as much gain as possible for himself as well as for his employer. To prevent the possibility of fraudulent deals between the lucky miner and his neighbour who had pitched upon a capricious lode, contracts were drawn up according to a printed set of regulations to which all miners had to agree; for instance there was a fixed scale of fines for the non-fulfilment of a contract. But the actual contract was the result of direct bargaining. On an appointed day the men would assemble at the company office, having previously inspected the work that was to be auctioned, and bid against each other for the pitches. The mine captains, who had already assessed their mineral value, did not necessarily accept either the highest or the lowest bid but usually those that were reasonably economical. Not less important and significant was the method of payment for what was known as "tut-work", by which men were paid so much a fathom for the sinking of shafts and the driving of levels. Thus both management and men were mutually interested in the tasks of excavation and extraction; it was claimed that this system provided a natural and easy discipline for controlling a large body of workers; and it certainly promoted in the miner caution, economy, forethought and judgment. All the blasting powder, cartridges and candles he had to buy from the company and the cost of these would be deducted from the value of his ore; so he needed to exercise care in his use of them. Sometimes before the monthly pay-day he would be overdrawn and then the company would advance him a loan, the repayment of which again tended to make him careful in his budgeting. Their unique partnership in the management side of the industry meant that in a way they behaved as small capitalists, entrepreneurs and shareholders in an underworld that would frighten the layman, where life, liberty and the pursuit of happiness depended almost entirely on their own initiative, coolness and skill.

Transported to America by the Cousin Jacks, the system of individual bargaining gave them a strength and command that other groups had to find by political means if they wanted power in the community. Political awareness has never been a marked characteristic of the Cornish because of the peculiarities of their occupations. Mining, farming and fishing by their very nature demand a withdrawal from the crowd or group, where political ideas are generated, to the field, the open sea and the underground passage where men gather only in twos and threes, making them reflective rather than revolutionary, and content to play a minor role in local government and in the development of free masonry. Many indeed were elected sheriffs. But on the whole their political loyalties were never crystal clear, not on account of indifference or irresponsibility, but because of their paradoxical position in American society. Industrially they tended to be reactionary because of their own vested interests in mining and the fact that they provided the captains; socially they tended to be radical because they were of Methodist rather than Anglican persuasion; but, when called upon to vote, they declared themselves Republican, some bearing arms in the Civil War and a few deserting after drawing their first pay and uniform. Because of their experiences of poverty and hunger in Cornwall they might have been socialists and violent leftists but Methodism directed their discontents into more constructive channels. Through the warmth and fervour of its church services and then through the intellectual disciplines of the weekly class, Methodism unmasked the folly of ever expecting that the New Jerusalem could be found in Cornwall where a conservative Anglican Church preached that man is by nature bad, that religious institutions alone make him good, and that emigration was the act of a coward trying to avoid the inescapable facts of human existence. Unemployment and near starvation pushed the miner in Cornwall out of mine, cottage and family. New prospects and opportunities pulled him to America, but it was Methodism that provided the urge, inspiration and resolution by which he became successful in an enterprise where there were many failures.

It is not surprising then that, wherever the Cousin Jacks pitched their tents, in Wisconsin or California, Michigan or Arizona, the first permanent buildings they erected were the Methodist Church and the Sunday School. Their repeated avowal never to work on Sundays, their "faith" teas, their hymn and "curl" or carol singing, their Bible study in the weekly class, all contributed to bringing to frontier settlements the disciplines of a social and religious code where none before existed. It was not exceptional for a Cousin Jack to be superintendent of both a mine and the Sunday School; nor for a tributer on the Monday morning shift to have preached the night before. No doubt his oratorical performance appeared rough and unpolished but so did almost every expression of Cornish culture. They carried with them to America no outstanding body of native literature (the lack of it had killed their ancient language long ago); they could not boast of a single national poet; the harsh, tough granite and the exhaustion of their woodlands for smelting and firing prevented them from developing a folk art; and their continual crawling in the stopes of their own remote world ill fitted their bodies for the country dance. Yet they could sing in harmony naturally, while a fluency that expressed itself in the "plod" or tall tale was to become the very stuff of entertainment in the mining camps. Their preference for the instruments of the brass and silver band grew naturally out of their own industrial landscape: the tuba, cornet, trumpet and drum were more in tune with the heavy beat of the stamps that crushed the ore and the din of the drinking saloon than the violin, 'cello and harp; and the hand that wielded the pick and sharpened the drills was perhaps insensitive to the string. Their physique was better styled for demonstrations of their prowess in the arts of wrestling and boxing (every Cornishman remembers their blacksmith champion, Bob Fitzsimmons of Helston, who won three world titles at different weights), or at drilling contests, single or double-jack. They are part of a culture that has survived because its very roughness gave it a resilience and a durability that enabled it to be transplanted in an alien soil where it easily took root. It is not without significance that at Christmas the old Cornish carols are still sung on the steps of the Union newspaper offices at Grass Valley, California; that every summer the melodies of Wesley's hymns float around the deserted shafts on the Keweenaw Peninsula in Michigan; that Cornish flautists and trombonists provided the wind section for the first concerts given by the San Francisco Symphony Orchestra; and that at Central City the splendid singing tradition of its Teller Opera House begins with the Cornish choirs from the silver mines in Gregory Gulch.

For the emigrant from Cornwall, adjustment to the harshness of life on the mining frontier was never quite the traumatic experience it was for other nationalities, emphasizing their sense of isolation and estrangement. A mine is much the same the world over as a harbour is to the homing fisherman. The Cornish were both miners and fishermen; and indeed they used the same terminology for both pursuits, for a mine was under a captain and its depth was measured in fathoms. Ranging far over the uncharted surface of the ocean and into the untapped depths of the earth, they were unbowed before the terrors of either and so acquired, whether in the eye of the storm or at the turning of the earth's axis on the "grave-yard" shift, a metaphysical unity with the universe that left them meditative and reflective; it seemed as if the whole world was their province, as indeed their history has proved. None is more cosmopolitan or more attuned to nature than the Cousin Jack for he has worked at the bottom of every hole in the earth's crust; the feeling of being near its very centre gave him an assurance and a confidence that frontier violence could never dislodge. Out of it sprang an easy accommodation to the world that helped him to tolerate other races; and so the Cousin Jacks became the catalysts of a new society, creating the conditions of wealth for the birth of another nation, though they were rarely wealthy themselves. With pick and gad propping up the palings of the frontier they "smoked pipe" and drank like leviathans, at one and the same time the most and least American of all the emigrants for they could don and doff the costume of a new people almost at will. Chords of remembrance are still struck every year across the Atlantic. A Southern Californian Cornish Association meets regularly at Los Angeles. A Searle returns to St. Austell to tell the tale of the American blood that ran in his veins at birth, for his mother was a King whose four brothers successfully farmed sheep in Laramie. Another American Cornishman breaks into business arranging the sea voyage home for the descendants of emigrants. And Cornish newspapers each week carry items of information about its "exiles".

The story of Cousin Jacks and their Jinnies on the continent of North America begins in the shallow lead mines of Wisconsin in the 1830's, and closes in the shadows of the vast copper amphitheatres of Arizona at the century's turning. Its heroes and heroines are ordinary men and women who would be the first to admit that their individual lives have no special claim to be remembered or recorded. Yet from the family sagas emerge epics of bravery and fortitude, bad luck and hardship, disappointment and despair; of death by cholera, typhoid, silicosis or salivation, or by physical annihilation. Some of their names are cut in granite tombstones or carved in bleaching wood; others, collected by the Women's Institutes of Cornwall, are recorded in Truro Cathedral. Yet these names were once flesh, nerve, bone and sinew that together shaped a personality. One, that of Dick Bullock of east Cornwall and South Dakota, passes into fiction as Deadwood Dick, the mythical hero of the dime novel. Another, that of the singing blacksmith Richard Jose of Lanner, wins fame beyond the countries of his birth and choice. But the majority went about their daily business, content to excel in their crafts, active in their masonic orders, singing in their choirs, and doing their best for their children because they were "lent" to them by God. America made them and they helped to make America.

II

WISCONSIN AND ITS CORNISH "BADGERS"

Of all the Cornish settlements in America those in Wisconsin are the earliest in time and the most enduring, faithfully reproducing the traditional occupation patterns of the homeland and superimposing them on an older pattern that derived from the years when Wisconsin was part of New France. The miners, farmers and stone-masons beheld a land of green woods and the blue waters of a thousand miles of rivers that stream into the Mississippi: of dogwood and wild plum; and of brown bluffs and rolling pastures. But they also entered a community that was both settled and politically mature, conferring on them rights denied to those left behind in Cornwall for more than two generations: they could vote after a simple oath of allegiance; their wives could control their own property; and small homesteads were exempt from seizure for debt.[1]

Any account of the Cornish in Wisconsin must begin with an acknowledgement to its first American historian, Louis Albert Copeland, who in 1960 was living in retirement at Duarte, California, and whose one regret was that he had never visited Cornwall. In *Wisconsin Historical Collections* (vol. 14) of 1898 he published *The Cornish in South-West Wisconsin*, an excellent study based on information he assembled from pioneers who were still alive or their children, from reports in *The Galena Gazette*, and from census returns; and it is the starting point for all subsequent investigations.

It was the lead regions around Galena that first drew the Cornish to Wisconsin, the industry being concentrated on both sides of the Fever River, some six miles from its junction with the Mississippi and 400 miles above St. Louis. Here, where the land was soon to be pitted, as it is now, with mounds of spoil as if untold hordes of badgers had been at work thus giving to Wisconsin the more familiar name of the Badger State, the first burrower in the earth to arrive from Cornwall was Francis Clyma. Born in Perranzabuloe in 1792 he reached the diggings at Galena in 1827 where his wife, Frances Maynard of St. Ewe, joined him, only to find that she had to be placed under the protection of Ferguson's Fort while he built the log cabin, so serious was

[1] R. A. Billington, *Western Expansion* (New York: The MacMillan Company 1949), p. 309.

28

the Indian menace. Bearing arms against Huron and Winnebago was an experience that these early Cornish had not expected, and several under the stress of battle discovered hitherto unsuspected qualities.

Those who came in their hundreds from Camborne in the 1830's (the figure was to swell to about 5,000 by 1850) were, after the tedious journey by way of Quebec, Detroit or New Orleans, in no physical shape to face arrows and tomahawks in the slit-trenches of their mines. Yet one of them, Edward James, held office under Colonel Henry Dodge, hero of the Black Hawk War, a miner himself and the mountain man who had been Marshal of Missouri Territory at the age of twenty-two. James was twenty-six when he reached Mineral Point across ocean and river from Camborne and not thirty when he fought along-side Dodge at the battle of Bad Axe. "Brave, intellectual but rather restless", according to Copeland, he seems to have been well liked by Dodge for they went mining together after the war until 1836 when Dodge was sworn in as the first Governor of the Territory of Wisconsin. One of his first acts was to make James his private secretary, but how efficient he proved with pen and paper is not disclosed. He was probably a tougher fighter in the field of public order for in 1837 Dodge made him a Marshal, a position he held for only four years, for the fever of the wanderer gripped him again among the cattle and the cotton of Missouri and he died at St. Louis in 1845. As for the remainder of these early arrivals information is very scanty. Francis Vivian, at Mineral Point with his family in 1832, served as a soldier in the Black Hawk War, and William Phillips founded the first Methodist Episcopal Church there. And in spite of his careful case-history of the Cornish we would know no more of the family of Copeland but for a correspondence with him over the years that has unearthed detail that might well sketch the pattern of settlement of many other families.

His connection with Shullsburg, the town that perpetuates the name of Jesse Shull, an Indian trader who worked for the fur empire of Jacob Astor, begins with Samuel Richards from Treswithian Downs and the brown moorland that sweeps to the mines of Camborne. In 1848 he persuaded his brother George and his wife to join him in Wisconsin. They left with no regrets since of Cornwall itself they knew little and the Land's End was only a name, for though it was only 25 miles away they had never seen it, as it was "too far to walk there and back in a single day". Neither of them could read or write and their brother's letters had to be deciphered by a class leader of their Methodist Church. But their native ability that ran to waste in Cornwall was soon to be turned to good account in Wisconsin. He found no difficulty in procuring a lease on a percentage basis to dig for lead and, when he had made his first profits, his wife insisted that, as an insurance, he should buy forty acres and plant them with flax; within a year he had bought a further eighty acres and extended his sowing to include maize, wheat, oats, rye, barley and clover.

Adjoining the Richards' farm was that of the Copelands, one of whom had been mining in California until he returned to Shullsburg and married Jane Richards, one of the children of the Camborne pioneers. But no sooner was their first child, Louis, born than he carried them off to the West to run a store for miners at Humboldt Basin, only returning to Shullsburg when the Modoc Indians broke out of their reservation in 1878. But this time his Cornish wife, having seen enough of scalps, saloons and scavengers, insisted that her wanderer should set himself to farm on the profits of his mining store. Business man rather than farmer, he consolidated a 500-acre stock farm, continued to manage some enterprises in Oregon at a range of 2,000 miles, established a county bank in Shullsburg and became one of the wealthiest men in LaFayette County. Today the Copelands and the Richardses are best remembered for their town house of red brick with its landscaped gardens, fountains and wrought-iron fences, and for the opera house they built, now a hardware store. Nearby are the premises of the local newspaper, *The Pick and Gad*, for which Louis Copeland wrote many articles about the Cornish long after he had emigrated from Frederick in Polk County (where he was both judge and Sunday School teacher) to the roar of Los Angeles and the vice-presidency of the Lincoln Savings and Loan Association.

The Tregonnings and the Rules from Redruth likewise left their marks on Shullsburg, for they were a "God-fearing lot" say two of their grandchildren[1]. Samuel Tregonning had seen heroic service with Colonel C. H. Gratiot looking for copper in a howling wilderness of snow and ice on Lake Superior. Henry Rule had brought his family over from New York in a covered wagon, intending to make for California; but wisely stayed in Shullsburg and farmed. Staunch Methodists, they brought some semblance of sobriety to a mining town where the pastor wore his guns in chapel and the Cornish widow, Peggy Gregor, had been known to clear her saloon of rowdies by mounting a table with an armful of beer bottles. They encouraged all and sundry to submit themselves to the daily reading of the Bible, the weekly class, and the Sunday sermon and singing. All their working life was centred

---

[1] Information from Miss Carrie Tregonning of Shullsburg and Mrs. Carrie Webb of Janesville, both now deceased.

round the Centenary Methodist Church, founded in 1867, and still an imposing building with its slender white spire, its memorials to the Cornish dead and its reproduction of Holman Hunt's "The Light of the World". From here emanated a Christian charity and a sense of social purpose that have their place among the secular organizations such as the masonic fraternities; and their companion sororities, to which many a Cornish housewife belonged; and the Women's Literary Club, to which their grandchildren belonged[1]. One of the Rules trained the choir and led the congregation in its singing for more than sixty years; another, it was said, always kept the family Bible on the kitchen table for quick everyday reference. Few of them drank wine, not even at Christmas when they paraded the streets singing their "curls". Only on New Year's Day would they lapse into splendid abandonment at a "Tea Meeting" to raise funds for the minister's salary.

Over the prairies too roamed Prideaux from Camborne, all of them miners there; but business men and farmers in Wisconsin, presided over by Henry Prideaux who, with his wife Ann Treloar, was firing and clearing the tangled bush around Dodgeville in 1838. One son made a small fortune of 8,000 dollars in California and then bought a general store in Dodgeville. Another enlisted in the 32nd Wisconsin Infantry, served under Sherman in "his wonderful Atlanta campaign to the sea", prospered in the lumbering business at Mineral Point, was elected its Mayor four times, and built a new High School and improved the salaries of its teachers. James Prideaux left a Camborne torn by the agitations over the Reform Bills of 1832, steadily banked his profits from lead mines at Mineral Point, sent his parents the fare to join him, married a Cornish girl and, after her death, actually returned to Cornwall for a replacement. He was completely American, having served in the 2nd Wisconsin Infantry, and raised twelve children to work his farms.

Over and over again in Wisconsin the pattern of the miner-turned-farmer repeats itself, an adaptation that happened on such a scale nowhere else in America and is accounted for by the astonishing fertility of the soil above the mining ground, the gradual exhaustion of the lead lodes and the self-evident prosperity of Cornish farmers already in occupation. Samuel Treloar of Roseworthy piled a sod cabin for his family at Linden while scooping the shallow trenches for lead; and then

[1] It was founded about 1900 by the wife of a Methodist minister and was undenominational. Its objects were "to bring together women interested in intellectual culture and by organisation to strengthen individual effort". It was still active in 1959.

farmed. James and Ursula Hendy, from the smooth rounded hill of Pembro that looks down on the fishing port of Porthleven, were in Dodgeville in 1842 working a smelter with another Cornishman of the name of Cox who gave his name to a dip in the land that is still known as Cox Hollow, and then worked the fields with some Grenfels from Gulval. Around Janesville the Winns planted where they had once mined, provided the town with mayors for eighteen years, and recall that Jenkinsville close by took its name from a Cornish miller who built a grist mill on the banks of the river[1]. Marmaduke Trebilcock from Ludgvan mined in Shullsburg and then invested in eighty acres that are still farmed by his descendants. The Leans from Blisland, for instance, in 1845 were already settled on a quarter section of land near Palmyra and induced the Uglows, the Jollifes and the Hoopers to join them. Emmanuel Crapp from the wooded creeks of Gweek with his Helston wife "as soon as he could, bought a horse and a cow and saved up enough until they bought a farm near Potosi". Near Palmyra too were, and still are, the Trewyns, putting down roots as permanent and as tough as those of any juniper. At the Home Farm, two miles south of Oak Hill, six generations have ranged their cattle, organized their marketing and milk co-operatives, seen their children through universities, voted Republican, and worshipped as Methodists and Congregationalists.

Most of these Cornish farming families are better known to us than those of the miners since they were more settled; for folks who do not prowl the earth's surface freely lumber their basements with the raw material of history. Harlin G. Loomer, "born on the Gibbs Farm" near Whitewater in 1899, and today a prominent member of the Philadelphia City Planning Commission, is descended from a line of Cornish and Devon farmers from Camelford and North Petherwin that included one who won national fame for his work in China[2]. His great-grandfather was the Devonian John Baker born in 1809 of parents who were freeholders of the lands they farmed. Though their fortunes rose and fell with the uncertainty of the seasons, they hardly ever had to face the harrowing poverty that drove so many of their neighbours across the border to America. Baker's Cornish wife, Grace Bone, however, was his social superior, for her family had seen service in the Royalist household of the Dukes of Buckingham during the time of England's

---

[1] Jenkin's farm is still worked: and the Primitive Methodist Church, built on land he donated, continues as a place of worship.
[2] 'China's Friend in Need, John Earl Baker', *The Milwaukee Journal* (19 August 1957).

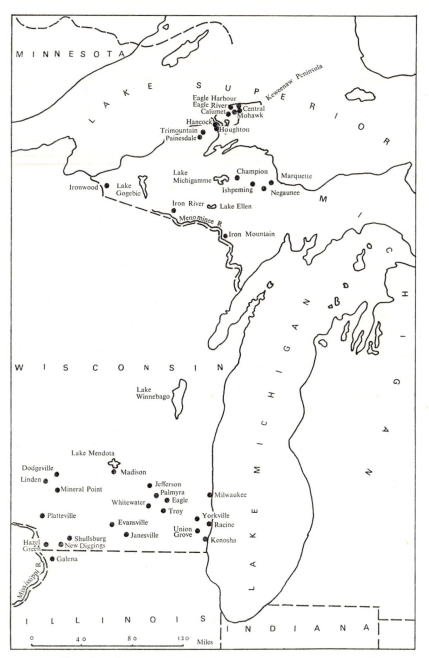

Cornish settlements in Northern Michigan and Southern Wisconsin

Civil War; indeed their Treludick Farm still bears the inscription "I & M.B. 1642" on an outbuilding, erected in the very first year of hostilities[1]. The marriage proved unpopular and, after years of humiliations, they decided to emigrate. It seems to have been a hasty decision for, although there were six children to consider, one less than two months old, they sailed at the worst time of the year, when ice-packs off the American coast made voyaging extremely hazardous and they reached Wisconsin when it was too late to plant crops. But their new home was well within the areas of settlement and far from the true frontier, with no Indians to fight, no ground to clear before the first sowing, and only a day's ride from Milwaukee.

The details of the voyage that began from Plymouth in April 1849 are contained in a diary that also includes notes which indicate that Baker had been in touch with settlers already in Wisconsin, and especially John Peardon of the township of Eagle, near Palmyra. He came from the same Devon parish as Baker and sheltered the new immigrants until a farm had been found for them. It was situated in the wooded valley of what is now known as the Kettle Moraine State Forest, 192 acres of "bur oak and white oak and living water" which he bought "for the sum of 310 sovreigns". One of several level tracts of the very fertile Little Prairie, it included a "dwelling house"; a log cabin with a loft, fire-place and a roof of split shingles that would "shed" the rain; 80 acres of "buck lot" that had already been cleared of trees and could be easily turned under with a breaking plough; and 112 acres that needed clearing. Gradually the wilderness of the moraine was subdued, the log cabin replaced by a house of stone, stock bought, and the first wheat crops reaped by a "grain cradle" such as was used in Cornwall. In a letter of 25 January 1851 to his brother William[2] to reveal the news of the birth of his first American son[3], he writes vividly and enthusiastically of his new life, urging Devon and Cornish farmers to follow him: "I and my family have not had one day's sickness never since we came here. Wm. Turner could do well here or any sober industrious Farmer that has two or three hundred pounds or any man that will work. The poor man that will work can get money just as fast as the farmer."

And work Baker did, struggling for ten years against a climate he did not fully understand. Season after season he fought a blight that settled unaccountably on the ripening fields, seriously contemplated

1 It has officially been designated as a building of historic interest.
2 This letter was discovered by John Earl Baker in Camelford, Cornwall, in 1945.
3 This was the seventh child and the first of three to be born in America.

returning to Cornwall or Devon, and then considered that perhaps his failures were a judgment from Heaven, as an entry in his diary for 1853 suggests: "I John Baker by grace of God do intend to totally abstain from all intoxicating drinks and from Sabbath breaking and from swareing". He seems to have been mindful of the dire consequences of succumbing to the temptations of the flesh and the devil, for he was by training a Methodist and ready to join any group of farmers nearby who would talk personal salvation, sing hymns and listen to his violin. At Little Prairie there was already a congregation of Bible Christians who, in 1851, organized themselves into the Methodist Episcopal Church, ministered by visiting circuit riders, and it was among them that Baker and his wife sought for a spiritual blessing on their efforts. But time was to prove that he was a better farmer than most, for he rotated his crops as he had done in Devon and Cornwall, whereas his American neighbours bled the soil dry and burned off the straw until it was "wheated out". By 1861 he was known locally as "Squire" Baker and his five sons were carving out an agricultural empire[1]. He had built a "country seat", planted orchards and increased his holding to 385 acres, of which at least 250 acres were earmarked for wheat just at the time when the Union troops required feeding. War prices rocketed the value of land and grain so that, by 1878, the Bakers owned 1,211 acres, worth 25,350 dollars.

Now they were not the first from the West Country to farm on a large scale in Wisconsin, for they were almost ten years in the wake of a remarkable group pioneered by John Foxwell of Mullion on the Lizard Peninsula. He was born in 1808, the first son and third child of William Foxwell and his third wife, Ann Harris. Like the Bakers and Bones, the Foxwells belonged to a small but significant Cornish squirarchy that might, in the fluidity and mobility of eighteenth century society, produce a Fellow of the Royal Society, an engineer, a chemist or a novelist, however humble their origins. William Foxwell

---

1 Harlin Loomer relates a remarkable case of surgical improvisation on the part of one "Doc" Salisbury who operated on Will Baker after he had been kicked in the head by a horse. His sole instruments were a needle and thread and a pair of tweezers made from a silver dollar: "After dipping the tweezers in boiling water, he proceeded to pick pieces of dirt and bits of bone out of the place in Will's head which had been punctured by a calk of the horse's shoe. When he had finished, an opening about the size of a silver half-dollar was left in the bone-case of Will's head. Again he ordered John to beat a silver dollar into a sheet as thin as paper. This too was boiled for a few minutes, then cooled to body temperature and fitted over the opening of Will's head, with shreds of skin drawn carefully over the wound and sewed together. Will recovered without further complications and carried this silver plate the rest of his life—a long one; with only the complaint that he developed a headache if he slept too long on the scarred side of his head". There is no mention of any anaesthetic, not even whiskey.

was already nearly fifty when his son was born, and a devout Methodist, for he had been converted by a travelling preacher in 1784 in the knowledge that his wife's grandfather had once sheltered Wesley on one of his many perilous missions to Cornwall. A class leader and local preacher, he was completely self-educated, learning from whatever books came his way; and no doubt this insatiable curiosity for knowledge he passed on to his son[1]. But he was also a competent stockman as we know from one of his text-books, John Rawlin's *The Complete Cow Doctor* of 1804, which is now in the possession of Dr. Alton Moyle of Madison, one of his descendants, not only lineally but also professionally. Deep in the Iowa prairie at Elgin lives the genealogist of this Mullion family, Thomas Foxwell, farmer, teacher and craftsman in wood, as were all the Foxwells before him. He has carved with exquisite care the altar rails in the Community Church at Illyria, where the Foxwell hosts are buried; and he has preserved a telescope and silver medal that once belonged to William Foxwell. It was through this telescope that on the night of 27 December 1807 he sighted from his Cornish farm the English frigate, the *Anson*, drifting towards the sands of Loe Bar, where she finally grounded; the medal was a public gesture for the part he played in the rescue of the crew and its military passengers[2].

William Foxwell's death in 1837 when his son John was almost thirty came at a time when the brown fog of the depression that was to last a decade and make the 1840's the "hungry forties" was curling its chill fingers from the sleeve of the Channel around the cliffs of Cornwall and creeping inland. Thoughts of emigrating were being considered by John Foxwell but 1837 was a bad year in America and the final decision was delayed until 1840. With a neighbour of the name of Thomas they shipped at Falmouth for the long sea route to Wisconsin by way of Quebec, the St. Lawrence and the Lakes and so to Racine, from where they walked to Yorkville over the vast untilled prairie. There he bought government land at 1.25 dollars an acre, established what is now known as the Perry Vyvyan farm and then, in September 1841, contrary to custom, married an American, Lucy

1   William Foxwell wrote an account of his life which later appeared in an obituary notice by his son-in-law, Samuel James, in the *Wesleyan Association Magazine* (July, August and September 1838). The facts are repeated in the *Foxwell Centennial Souvenir, 1841-1941*, compiled by H. C. Duckett of Burlington.
2   *The Commemorative Biographical Record of Racine and Kenosha Counties, Wisconsin* (Chicago: 1906), p. 102, wrongly attributes the medal for his gallantry in saving the crew and troops of the *Royal George*. Some of the guns of the *Anson* have recently been salvaged and may be seen in the harbour yard at Porthleven.

Philena Briggs, who could boast that her family had been in America since the seventeenth century and that her paternal grandfather had fought English troops in the War of Independence, had been born in Ohio in 1820, and had travelled with her parents in a wagon to Wisconsin when she was seventeen. Now she was just twenty-one and teaching in a log school-house that was also used as a church, ready to pioneer with John Foxwell in the log cabin he built on a second farm he had bought from the sale of the first. This home they called Lily Grove because of the profusion of wild lilies and apparently it suited her, for incredibly she turned the corner of the nineteenth century, looked back on the Civil War, and lived into World War One. She died in 1919, just missing her own century by four months.

The country around Yorkville was, as it is today, attractive beyond measure with the wide prairie lush with long grasses and wild flowers, the grazing ground for droves of cattle from the south. John Foxwell had no doubts at all that this was the place for the rest of his Cornish relatives, and thither they came in 1842; his sister Susan and her husband Thomas Moyle with their baby in arms; and his mother, by now in her sixties, who jokingly remarked that, since her husband William Foxwell was dead, there was no place in Cornwall for her to be buried: his two previous wives had been buried on each side of him and she had no wish to be laid to rest at his feet. The immigrant ship that left Falmouth on 4 April and reached New York on 19 May was the *Orient*, a hulk of an East Indiaman that was never fit to put to sea again. According to Thomas Moyle who kept a log of the voyage, John Leuty, the captain, "prayed long, loud and often on the way over, promising the Lord if He would save him this once more, he would never take the old tub to sea again". The burden of responsibility understandably weighed heavily on his professional conscience for his passengers to a new life numbered 182, among whom were forty mothers and seventy-seven children, their ages ranging from two months upward. Indeed one child was born on the way across and one mother died in the teeth of a squall, leaving the distracted husband to comfort their three children as best he could. They came from the hills and valleys from Camborne to the Lizard and Falmouth, from farm and mine, single and married, young and old, poor and not so poor, but all to endure the common frightening experience of the Atlantic under steerage conditions. By far the majority of the men were miners bound for the lead regions of Wisconsin but, on arrival, many went no further than Racine County and farmed, too tired to fare any further.

The destination of the Moyles, the Foxwells and the Jameses, already interlocked by marriage, was Southport (Kenosha today) and then inland by prairie schooner to Yorkville where John Foxwell was awaiting them with his American wife. He had already staked out farms for them around Caledonia, near the Root River, where he now bought for himself 160 acres from the profits on his sale of Lily Grove Farm. The census returns of 1850 for Caledonia show him enjoying a modest prosperity. His farm implements were worth 100 dollars; he owned two horses, three milch cows, two working oxen, eight "other cattle" and five pigs, the whole valued at 225 dollars. His crops were wheat, Indian corn, oats, Irish potatoes and buckwheat; he raked off 30 tons of hay for winter feed; and his annual butter production was 400 lbs. He moved only once more, back to Yorkville in 1856, where he purchased a small farm of improved land, and there he remained until his death in 1875. How much more valuable this farm was can be seen again from the census returns of 1860; his real estate was valued at 4,000 and his personal estate at 600 dollars; and the improved land of 149 acres commanded a market value of another 4,000 dollars. There is little reason to doubt that his relatives were not less successful.

From John, "the pioneer", who has been described as "a power among his neighbours for good, and one of the founders and a lifelong supporter of the church and society at Yorkville" has stemmed a Foxwell stock that, culturally, must be one of the richest in Wisconsin[1]. Twice, in 1914 and again in 1963, the clan has gathered to celebrate their historic moments in time, the first to honour "the pioneer", and the second to pay homage to the last surviving daughter of his brother Thomas Harris Foxwell who, in 1863, traded his farm in Caledonia with his brother-in-law, Walter Humphrey, for 360 acres near Elgin in Iowa. Here today his grandson and his wife guard the same acres and cherish their family trees with a care not less than that of their forebears who ploughed their straight furrows.

The decision of his Mullion relatives to join John Foxwell only came after the most careful consideration, as two important letters to him from his sister Anna Maria and her husband Samuel James suggest[2]. The first, sent from their home at Trelan, just outside St. Keverne, and dated 17 October 1840, reveals that "emigration thoughts for the present we fear must stand aloof at least as far as we are concerned"

---

[1] *The Commemorative Biographical Record etc.*, p. 102. One of his descendants is the distinguished American Orientalist, William Foxwell Albright.

[2] In the possession of Mrs. Susan Thomas of Los Altos, California, to whom I am grateful for permission to quote. For further information about the Jameses family, see section on Oregon, p. 209 *et seq.*.

because of a promising speculation in the "sale of marl", even though "we are racking our brains, racking our limbs, racking our lands in order to make them turn to more account than the Creator ever intended them for". But within a year the die was cast, not on account of poor harvests for "potatoes are everywhere a most abundant crop", but because of the rising curve of taxation due to a foreign policy of which they disapproved: "The truth is Russia, England and France are trying to outgeneral each other and obtain possession of the East . . . . In consequence of our warlike attitude we have a fresh accession of taxes. All the assessed taxes are raised 10% and the probability is further impositions must take place". So on 23 October 1841 Samuel James writes to John Foxwell: "I am at last determined by the Divine assistance to wind up my affairs here and set my face towards the land of promise. I should like to settle near you if possible on which account I would be willing to forego some qualities in an Estate which I would otherwise deem to be desirable. I should like if such might be obtainable to have a gently inclined surface with here and there a living spring. Let us hear from you immediately for what sum you can purchase and how we can remit you the purchase money".

There can be no doubt that the exodus that trailed from the Lizard and Camborne in 1841 gathered momentum throughout the "hungry forties" due to these first emigrants who sent back glowing reports of their new homes to fan the flagging spirits of the timid and the uncertain. Mary Carter Thomas, the wife of Thomas Foxwell, brother of John "the pioneer", writes from Caledonia on 27 February 1843 to her sister in Mullion[1]: "as to the state of society here it is good they are in general wellbehaved, there are no locks, bars, nor to the house nor do they need any, the people go away all hands for a week or two to camp meetings sometimes turn the calfs on the cows and put two pieces of sticks cross ways before the door to show they are gone, when they return they find things as they left them. I believe those who feel themselves hard done up in England would like this country." In June 1842 Ann Foxwell writes to Mary Ann Gartrell to persuade her to emigrate to her new land of "beautiful Horses and Bullocks, very fine rich pastures: beautiful tall groves and wild strawberries in abundance". And since she cannot tear herself away from Cornwall, Matilda Foxwell two years later adds: "I often think on the pleasant hours we spent together when at home but I feel no desire to return to a country oppressed with taxes".

---

[1] Quoted by kind permission of Mrs. Reed Johns of Porthleven, Cornwall, who is a descendant of Mary Carter Thomas.

So to Wisconsin thence they came in their hundreds. The Goldsworthys farmed 120 acres at Yorkville; grandfather thundered from the pulpit on Sundays and from the magistrate's bench on weekdays, and father "gave his aid and moral support to the loyal party during the war". Thomas Dale, by his "frugality and industry", tended 450 acres during the time when his brother was fighting as a Colonel in the 2nd Wisconsin Cavalry. Thomas Martin with his statutory 120 acres was "a saving hard-working man" and his wife "a faithful helpmeet". Samuel Skewes successfully combined the crafts of farmer, merchant, carpenter and builder, while the Primitive Methodist Vyvyans and the Episcopalian Luggs from Ruan Major cleared the prairie for their future fine farms[1].

One of the most thoroughly documented records of these Lizard families, that of the Moyles, with their affinities with the Foxwells, Jameses, Shephards, Georges and Harrises, is the work of their genealogist, Dr. Alton Moyle. His excellent account[2] begins with the forefathers of them all, Thomas and the farrier John Moyle of Constantine, who have so influenced the sons and grandsons that every generation has produced at least one modern equivalent of the farrier: the veterinary surgeon or, as he is termed in America, the veterinarian. Both Alton Moyle and his father follow this profession and indeed ride in the footprints of the emigrant Thomas along the same trails to the same farms. They own some of the medical books he brought from Cornwall, a contrivance for making pills for sick cattle, and the account book of the manor of Chiverton, the spaces in which he had utilised for his journal, so scarce was the supply of paper.

Thomas Moyle had married Susan Foxwell, the sister of John "the pioneer", in 1838, his first wife, Elizabeth Jose, having died the previous year. Like the Foxwells he belonged to the self-educated Cornish gentry that had defected from the Church of England for the warmer faith of Wesley and, as in the case of Samuel James, his library was his university, for only Anglicans were permitted to study at Oxford or Cambridge. Healing sick animals was his occupation, reading was his hobby and preaching was the expression of his faith. His reasons for leaving Constantine were substantially the same as those that influenced Foxwell and James, though the slight sketch of him in *A Portrait and Biographical Album of Racine and Kenosha Counties* (Chicago: 1892) suggests one other: "Thomas came of a respected and

1 *The Commemorative Biographical Record etc.* pp. 397, 453, 516.
2 Alton I. Moyle, 'The American Descendants of John Moyle, Farrier, of Constantine in Cornwall, England', *The Detroit Society for Genealogical Research Magazine* (Vol. 26, Nos. 1, 2, 3, 4 1963).

well read English family of that county (i.e. Cornwall), who had long been admirers of America's free institutions, hence he was thoroughly Americanised before reaching this country".

However, his importance lies in his ability to observe, to study and to record as parts of the daily business of living and earning a living. At Yorkville, as the census returns show, the graph of his investments curved upwards as steadily as that of his brother-in-law. But perhaps he was very much busier for, as well as growing crops, he was in constant demand up and down Racine County as the farmers' veterinarian, as well as treasurer, clerk and assessor of taxes to the town council. In his working journal he sets down with meticulous care details about service on the Grand Jury at Racine; pruning Mr. Lory's apple trees[1]; auditing his neighbours' accounts; attending the Kellog District School "celebration"; raising his house on runners and hauling it to another position with twenty-one yoke of oxen; burying his friends the Tamblyns; and stoning a well. But on 4 November 1868 he notes: "at Racine to Doctor", and this was his last entry for he died three weeks later, a Racine newspaper commenting appropriately: "Mr. Moyle has not left his equal in the neighbourhood".

For a quarter of a century the disciplined hands and eyes of Thomas Moyle were a civilising power to be reckoned with in the settlement of Yorkville by the Cornish, for a deep religious faith stood at the very centre of his public and private life. No mean student of theology himself—his library at Constantine was well stocked with the standard works and he took them to Wisconsin—one of his first tasks at Yorkville had been to establish the Wesleyan Methodist Association, a move that was welcomed by the few Cornish already there. One of them was Elizabeth Shephard, a traveller on the *Orient* in 1841 with her husband, Hannibal, two children and another to be born almost as soon as she stepped out of the ox-wagon. She too kept a journal[2] and in November 1841 laments: "I sensibly feel and my husband also feels that we are as few sheep without a shepherd, having no class meeting or prayer meeting, but we hope for better days". A year later she rejoices: "This year has brought some local preachers and class leaders from my native land. How rejoiced I was to see them, and presented my infant to be baptized. Mr. Hancock performed the ceremony calling his name Thomas Foxwell Shephard". Hancock was one of the lay preachers on the *Orient* and there is no doubt that one of the class leaders was Thomas Moyle, whose moral strength soon began to make itself felt

[1]   See Letta Lory Shepherd, *The Lorys of Cornwall* (Ann Arbor: 1962).
[2]   *The Foxwell Centennial Souvenir*, 1841-1941.

in the class meetings held in the home of Elizabeth Shephard and that of her consumptive brother-in-law, William Thomas. It was the latter's death in January 1844 and that of Hancock shortly afterwards that led to the idea of building a church, as Susan Foxwell recalls: "A young man, beloved by all, was taken from us. He did not forget, while living, the God who brought him safely across the ocean. He desired to see a little chapel erected." So, in March 1844, a meeting was held in the home of their class leader, Hannibal Shephard, to decide upon the building of a "meeting house" for their religious community that they called the Wesleyan Methodist Association of Yorkville.[1]

The church was small, no more than 28 ft long, 22 ft wide and 10 ft high, with walls of stone as high as the sills of the windows, and the rest of clay and straw, the actual building being assigned to Adam Hay, Hannibal Lugg, David Chalmers, Hannibal Shephard and Samuel Skewes, and the raising of money to Thomas Moyle and Jacob Lory. When it was finished we do not know but the founders seem to have been forced to fight local apathy and indifference as Elizabeth Shephard records in 1848: "I feel to mourn the low state of religion here compared with my native land. Oh that the Lord would pour out His Spirit upon the place". And three years later, after she has heard the funeral sermon preached over the dead body of her husband, the position has not improved: "Trials seems to await the society. Father forgive. The minister has left. My prayer is, 'May God convert his soul' ".

By 1856, however, their difficulties seem to have vanished for the church was then found to be too small for its growing congregation and a "mass meeting" was held to decide on improvements. Every year there are fresh elections to the Board of Trustees, and every year with unfailing regularity the Georges, Luggs, Moyles, Bosustows, Roskilleys and Iveys take up their positions and discharge their responsibilities. In 1865 they make provision for spring stops for the windows, web straps to lower coffins into their graves, and a plank road "extended 6 ft clear of the Church on the east side so as to allow two teams to drive up at the same time". And in 1902 they hang a bell in the tower.

Today the church has disappeared and on its site stands a memorial to the Yorkville dead of two world wars. It vanished in a swamp of argument and recrimination when, in 1914, a committee found it could

---

1 Account based on the Minutes of the Board of Trustees, in the possession of Mrs. Susan Thomas, who is herself a descendant of Thomas Moyle.

not proceed "to erect a new church on the old site under the present self-electing, self-perpetuating Board of Trustees" and called for resignations and elections "by the people". An emergency meeting was called "because of the Church having been broken into and the furnishings thereof stolen by the Methodist Episcopal people"; and, sad to relate, Edward Skewes and John Hay were accused of the desecration. A new church stands aloof from the disputes of the old. With time and progress continents foreshorten and oceans narrow, but Yorkville and Mullion, Caledonia and St. Keverne, Big Bend and Constantine change but little. In the small corner of south-east Wisconsin where the prairie comes down to meet the lake, the Cornish touch is still apparent. Yorkville, loud with bees on the hot summer air, may not have the granite cottages of Mineral Point to display its Cornish ancestry but, where the roadway railings are white, the barns a brilliant red and the silos a dazzling silver, the old frame houses of the "forty-twoers" still stand.

All their American history begins at this point off the road where cluster the homestead of pioneer John Foxwell and the Moyle and Shephard farms, now occupied by a great-grandson of Elizabeth Foxwell Shephard who possesses a set of Cornish jack-planes, a Cornish writing desk and some letters from Predannack on the Lizard, dated 1865, while in the garden cherries ripen on trees grown from seed sent from Cornwall. But it is in the old burying ground that the Cornish lie in all their abundance among five generations of Moyles and six of the Foxwells and the Skewes. If there is pride of place among the dead, even though death be the great leveller, it should go to John Foxwell as his monument proclaims, "born in the Year of our Lord 1808, died March 20, 1882. a native of Mullion parish, Cornwall, emigrated to Wisconsin, the Cornish pioneer"; to his mother who closed her eyes for the last time 3,000 miles away from her husband; and to William Thomas, the "well-beloved", the first to be buried there, in 1844.

In their venture into the unknown their first real fear had been that of the Atlantic crossing, a voyage of six weeks for which they had to provide their own food, the captain being responsible for only water. Thomas Moyle gives a vivid picture of conditions aboard the *Orient*, in spite of his own sickness, with a humour that is almost a diversion as the vessel crawls along at two knots an hour. With "great valleys and trenches" in the sea, all take to their beds, no fires can be lit in the caboose, the children start colds, the foremast is swept overboard, tangled ropes strew the decks, no water can be served until nightfall, and the passengers are "very much dissatisfied". But he is thankful that they are not already food for the fish, praises God that He has "prospered our way so far", and is "determined to be more God-like himself". This should not have been too difficult for the passengers included thirty-one Methodists, sixteen Associationists, and eight Bible Christians, though Moyle is not certain how close these Cornish are to God. Sermons are preached almost every day, but Moyle reserves judgment, for "many words darken counsel". But there are more pressing problems to solve for on 22 April he records: "Called by Captain to bleed one of the sailors. Captain gave me a rusty lance. I told him I could work better with my own tools". So the Lizard farrier lances the mate's thumb twice, "extracts tooth for Miss Edwards", and attends the sick in the fetid atmosphere of the fo'castle. And then, when half-way across, one of the mothers dies: "Squally-N.W. by W. At breakfast time a woman from Camborne, Mrs. Thomas, died. Left three children. At eventide sails were furled and ship hove-to while funeral lasted. At 5 o'clock the body of her whose spirit had gone to glory was consigned to the great deep wrapped in canvas with two bags of ballast". This was a sobering experience, for Moyle notes, with some anxiety, that the voyage had done little to demonstrate that man was by nature good. On the contrary, "find every man's hand against neighbour, and even friend, if anything is to be got thereby. Every man serves his own end even when to the disadvantage of others". But even so, he never regretted his decision to leave Cornwall: "Thinking of home. Thought much of Sunday School tea gathering. Would like to be with them. But satisfied with present prospects".

The routes from New York were many and devious. Some American pioneers continued to use the three oldest trails, the National Road, Forbes' Road or the Catskill Turnpike, then down the Ohio to its confluence with the Mississippi and northwards up river to Galena, a journey of at least 2,000 miles. But those coming from Cornwall took advantage of the Erie Canal, opened to traffic in 1825 and linking the Hudson River with Lake Erie, for it was reasonably cheap though slow. The Moyles, after being cooped up in the *Orient* for six weeks positively welcomed the leisurely canal journey for they were able "to get acquainted with the animals and the vegetation of their homeland". But Thomas Harris Foxwell, Thomas James and Walter Humphry, returning in 1849 from a visit to Cornwall, chose a most unfortunate time to make this journey. They found New York in the grip of a severe outbreak of cholera which stayed with them the ten insufferably hot days to Buffalo, and even accompanied them on the lake steamer. Many of the passengers were buried on the banks of the canal, and the worst fears of Thomas Foxwell, that America was a place where only the toughest could survive, seemed only too true when his nephew, Thomas James, was stricken with the disease aboard the steamer *Key Stone State*.

A third way to Wisconsin, favoured by the earliest of the Cornish pioneers, like Francis Vivian, was through New Orleans and up the Mississippi. An account of this passage has been left by James Skewis of Camborne who made the journey when he was twenty-three with his mother to join his father and brother in Shullsburg. There were other Cornish in the party that left Liverpool in January 1849 on the *Mary Ward*, and in New Orleans one of them, "one of our most robust farmers by name Pengilley" died of cholera; a shock that was not lessened by their first sight of negroes: "Slaves were sold in the auction market like horses but to us it looked different for those negro people had sense and reason and their pain in being sold and separated from loved ones was heart-rending". Once riverborne and moving north, however, the young miner was captivated and impressed by his novel surroundings where "the river had a great many turnings and beautiful trees and the houses of the slave-owners looked very showey and the little huts of the slaves made it look like a small village and the boats had to put in for wood then the negroes would by great torch light if by night load up the boat for another trip". At St. Louis he changed boats for the twelve days' climb up-stream to Galena, which he admired because "less negroes, more white people", and from there joined his father at Shullsburg. The reunited family vacationed for a few months at Dodgeville "where we had lots of cousins by name Hoskins, Pearce and others", and then returned to Shullsburg, James Skewis working in the lead mines "mostly on the Townsend land".[1]

Canada was another point of entry, for there was a fairly regular service of boats taking out unemployed labourers from England for resettlement by the Poor Law Commissioners; it was the cheapest route because the sea crossing was the shortest, but it was also the most dangerous because of ice, fog and strong headwinds. John Baker of Camelford chose this route in 1849 and has described their troubles off Newfoundland: "From the 23 to 30 (of April) the wind and weather hath been so changeable and I have been so much occupied in cooking on deck and room so small between decks and so many children to attend to that I have not kept a daily journal. I have suffered much

---

1 Information kindly supplied by Mrs. N. Hiday of Salem, Oregon. See also the section on California, p. 60.

while on deck cooking; the weather still cold it took great effect on my bowels and was ill, I think I never was worse in my life." They were attempting the southern passage through Cabot Strait into the Gulf of St. Lawrence but made little progress. St. Paul's Island came into view on 4 May, but a week later it was still only in sight, so rough was the wind "right before us". Under these trying circumstances their morale seemed no better than that of Moyle's companions: "We have many miserable people on board to contend with we have old John Hawe on one side of us, he is a shoemaker of Stratton, they are the biggest heathen I ever saw". But they were lucky to be alive for on 14 May: "The pilot came on board and told us there were about 150 vessels gone up before us and also said one of the capts told him that there were two ships lost in the ice he saw the ship sink the vessel loaded with passengers so they all had a watery grave". But this was the end of their tribulations for, after four days of pleasant steaming across the Lakes and two days of creaking in a shaking and springless lead wagon, they reached Melendy's Prairie, nine weeks and three days after leaving Plymouth; and the total cost was less than £25.

Along these shipping lanes the Cornish came to Wisconsin, the farmers to settle in the east where Palmyra, Whitewater and Yorkville look down upon the lake, and the miners in the west to Galena, Dodge-ville, Mineral Point, Linden, Hazel Green, Shullsburg and Platteville. Today these lead mining settlements are part of the "dairyland of America", where blue glass-lined silos raise their cylindrical towers from a depth of thirty feet below ground, though here and there on the old wagon road from Galena to Shullsburg they are rivalled by the giant beehives of stones taken from the mines, and by the drilling rigs of the new prospectors. Shullsburg unrolls a wealth of virtue in streets which are called Mercy, Goodness, Judgment, Wisdom, Peace and Truth, just as in the Cornish fishing town of St. Ives a pilgrim's progress is measured by Virgin Street and Teetotal Street. But the links with Cornwall fracture as the town shrinks. Louis Copeland no longer writes his articles for *The Pick and Gad*, and his Opera House is a forlorn shadow of its former splendour. Only the Methodist church with the clean lines of its white spire tells until time shall end, in memorial book or stained-glass window, the names of its Cornish worshippers: Rowe, Trebilcock, Tangye, Shephard, Glendenning, Kittoe, Rule, Skewes, Stephens, Berryman, Odgers, Tregonning, Sincock, Trewarthas, Trestrail, Tregloan, Hancock, George and Oates, and the Richards who mated with the Copelands.

The size of Wisconsin's Cornish population is difficult to assess.

Copeland examined the census returns for 1850 and calculated that, out of a total population of just over 9,000 in the mining regions of Mineral Point, Dodgeville, Hazel Green, Linden and Shullsburg, about half were Cornish. To these he added the mining settlements in northwest Illinois around Galena and arrived at a final total of not less than 7,000. If to these be further added Cornish farmers in the east, 8,000 seems a reasonable figure for the whole state. Nevertheless it can only be an approximation for in the 1850's many of the Cornish departed for California and about three-quarters of these returned to Wisconsin to begin life again as farmers. Yet Copeland reckoned that in 1898, the year when he completed his survey, at least 10,000 pure-blooded Cornish lived in the lead areas, most of them of mining stock. So it is that the Cornish miner takes place of honour on the Great Seal of Wisconsin.

The contribution of the Cornish to the techniques of mining was timely, for the only experience in Wisconsin, prior to 1823 when the Government granted the first leases, belonged to Indian squaws. Their methods were crude, for they neither sank shafts nor used gunpowder, merely drifting into the side of a hill. When they came upon a cap rock they simply kindled a fire beneath it and waited for the heat to split it, the final cracking being achieved by the application of cold water. They smelted the lead in small make-shift furnaces, built of loose stones and sand on the side of a bank of earth, similar to the early blowing-houses of the Cornish. By this process, however, the Indian women extracted more lead than the Americans, who made the mistake of applying too much heat in the early stages. But both Indian and American abandoned a lease when they hit rock, for they did not know how to follow a vein. It was at this point that the Cornishman began, delighted no doubt to find that he needed little or no capital, for the price of the lease was a percentage of the ore raised. Everything, of course, depended on his initiative at the diggings, wise speculation in land and careful household management by his womenfolk. But for the most part he was his own entrepreneur, he sensed and enjoyed his new independence and so he planned and built to stay.

Some of their solid houses, cut from local limestone, are still to be found in Mineral Point on Shakerag Street, so named from a custom of the Cornish women when they signalled their men from work to a meal. Three of them, *Trelawney, Polperro* and *Pendarvis* have been so well restored by Edgar Hellum and Robert Neal, whose Cornish grandmother emigrated to Wisconsin in 1847, that they have been scheduled as historic buildings by the Department of the Interior.

With their landscaped gardens, stone courtyards and gates of wrought-iron they are a living memorial to the first miner who, adept at cutting granite, found the Galena limestone hardly less difficult in that it required careful handling and an expert knowledge of the seams and formation to prevent it cracking. And further down the same street are other Cornish cottages still inhabited, with walls at least 18 in. thick, windows deeply sunk, and with two rooms up and two down with a staircase in the middle. Somewhat different, however, is the home built by a Gundry, a fine southern-type house, fronted with a veranda of white pillars and of red brick in spacious grounds of trim lawns and gravel paths, now the museum of the Mineral Point Historical Society. Other buildings resurrect the Cornish past: the hall of the Independent Order of Oddfellows; the Methodist Episcopal Church, founded in 1834 and splendid with the names of William Phillips, Andrew Rumphery and others; and a drugstore and funeral parlour managed by an Ivey and a Treweek.

The spell of Wisconsin on the Cornish was lasting, as well it might be, for they were the first national group to give it social and economic stability. Even as late as 1914 the trickle that had begun as a flood had not quite stopped: a Parsons in Whitewater; the Tredreas in Racine from St. Erth; and some Sleemans in Linden. By then, of course, the lead had all been exploited and all the Cornish of sensibility had become farmers, competing with the Germans, the Scandinavians and the Irish in the luxuriant valleys where the maize crackles as it grows in the stillness of a summer night. It is then that the Cornish of today meditate on the harshness of the life of yesterday, like John Treloar of Linden remembering that his grandmother lived in a sod cabin and that, as a boy, his mining life began at the Anaconda working a horse-whim. Sometimes the occasional letter comes to light that describes the anguish of parents parted from their children, like this one:

Hayle Mills Near St day
July 29th 1871

My Dear Son an Daughter:
I send you thes few Lines hopin the will find you all very well But i can not seay that i ham for i have not work for the Last 15 months nether do i think that i should Be able to work aney more for My Soul is Curist (crushed?)) in But thank god it is nor worst My Dear Son an daughter we Receved your kind and welcom Letter that Sent and was very glad an thankful to See you was Hall well i have Benn Expection one a Long time in Answer to the one that i Sent i was hurried . . .
My Dear Son an daughter we fel very thankful for the Lickness (photograph?) the are very hansum in our Eyes we should be very glad if we Could

kis them and you likewise and I hope you will be so kind as to send home all you likeness for Supose I should never See you no more in time But i hope to see you all in heaven My Dears I have writin to you with tears in my eyes My dayley Prayer is for my dear Chrildern that the will be Saved at last.

My Dears Por ould Mother is Stil alive But not able to do aney thing nor have not for the last 8 years and Ant Margery have been in the Bead the Last 8 Months an I do not Supose She will Ever Come down aney more. My Dear Chrildern I hope you hall try to find your way to Heaven Be good Chrildern and the Lord will bless you in this life and in that which is to Come you will have Everylastin' life with god in heven for ever My Dear Son daughter i Should Be verey glad if my other Sons and daughter in Law had the thought as you heave had i Should fell much Better for the do not Please me in there ways i have tould them how the ought to do But the Would not an i grive about it very much i do want to see my Chrildern do well Sister Anns husband is out to America in Place Call Plymouth Luzerne County Pna Box 63 North America his nane is Samuel Trefilcock.

My Dear Son i have Sent the West Briton So that will Leave know the news and i hope you will answer the letter and then i will Leave you know more news about the famley

So My Dears I give kinds Love to you hall and may god bless you hall give My kinds love to Por ould Mother Mother gives her kinds love to you hall and you must kiss all the dear Chrildern for us and hall you Sisters give there kinds love to you all Ant Margery Reed give her love to you So may god Bless you hall and i hope to meet you hall in heaven So no more at Present from you dear father an mother

William an Martha Painter.

So pathetic a letter, penned with much physical and emotional effort and little writing skill, though showing an astonishing power in some of its striking expressions and the almost liturgical chant of its invocation to the children, went its long way to William Painter of Gwennap and his wife Matilda Jane Francis of Kea, who had reached Mineral Point about 1850. The natural anxiety of parents, separated by immense distances from their children and ignorant of the world outside their own village, may have been worsened by their inability to express themselves in the written word, but on two counts at least the Painters were fortunate: doors were opened to new opportunities by which one great-grandson, after a brilliant university career, became a member of the American Consular Service; and in 1871, after twenty years, their children in Wisconsin were still writing home to Cornwall. For many parents the tragedy of emigration was that their sons in Wisconsin pursued their argosies afar in the Golden West, and disappeared without trace in the mountains and deserts of California.

III

CALIFORNIAN GOLD, MERCURY AND SILVER

To the memory of Samuel Uren, who was shot
the 11 of September in Califoyrnia 1875

SUCH is the startling entry in the slim pocket book of Matilda Uren of Redruth in which, ever since her son's departure in 1856, she carefully recorded the dates on which she received letters from him and on which she sent her replies. She had seen other Urens go and return: "John Tink and son left July 9, 1866 for Califoyrna and came home from there the 19 of November 1871". But for her own boy there was to be no homecoming; she never knew the name of the place where he fell nor the hand that fired the fatal shot; but in one simple sentence she etched with her pen the tragic impact on her life of the violence of the West[1].

A little over a hundred and fifty years ago, when England had repelled from its shores the military might of Napoleon and was extending its frontiers beyond India to Burma and Singapore, the golden land of California between the foothills of the Sierra Nevada and the Pacific Ocean was ruled by the imperial hand of Mexico that held within its palm a pastoral utopia for the dispossessed of Europe. Here, on the banks of the River of the Holy Sacrament, John Augustus Sutter from Switzerland had staked out his agricultural paradise of New Helvetia. In 1839, with the help of his Indian servants and labourers, he built a fort to protect it so that an immigrant from the East, after the ordeal of crossing the mountains, was sure of a night's accommodation within the safety of its stockade. Today his adobe fort is part of the city of Sacramento which his son laid out with precision and distinction for its role as a capital of culture and elegance, graced by twenty-eight parks and cooled by a profusion of royal palms and New England elm and ash. But in 1846 the tranquillity of both fort and city was shattered when the swashbuckling John Charles Frémont with his guide Kit Carson and their Americanos hoisted the flag of the Bear Republic at Sonoma, imprisoned Vallejo, the Mexican commandant, and abruptly terminated his country's Californian empire. Two years

[1] Information from Mrs. Marion Newton of Camborne, a great-granddaughter of Matilda Uren.

later, in the tailrace of Sutter's own saw mill on the south fork of the American River, James Marshall found gold; and the old Swiss gentleman was now to see his New Helvetia ripped apart by violence in the human earthquake that followed. A vast rabble of adventurers, careless of a land whose gold was the only harvest they desired, descended upon Sacramento in fleets of side-wheel paddle boats. Seven times the river protested and flooded as if in an effort to stem this unnatural tide but the placer miners still poured northwards to pillage and spoil, leaving behind on its banks mile upon mile of stones, once smooth and cool beneath the water but today arid and angled under sun and wind. Some met their death in the frantic search and were buried there; and in time Sutter's Fort rotted away from neglect and so did the pioneer cemetery at Sutter's Creek. In 1888 the Fort was rescued by the enterprise of the imaginative Native Sons of the Golden West, but four Cousin Jacks had to wait until 1963 for their earthly resurrection from the cemetery when a bulldozer, clawing at the earth to widen a road, bit into their headstones. Three are inscribed as belonging to "natives of Cornwall": John Scoble, died in 1874 at the age of 23; William Stenlake, died in 1868 at the age of 28; and Thomas Trudgeon, died in 1924 at the age of 70. The fourth, Alfred Venning, who died in 1899 at the age of 60, was a "native of the parish of St. Erth"[1].

Sacramento, Sonoma and Sutter's Fort mark the northern limits of the older and more gracious Spanish-Mexican culture. To their northeast it is all harsh and Saxon, a land of drifts (gaps between hills), flats (places where a tent could be pitched) and bars (the name for the river gravels that hold the gold). The modern highway, divided by pink flowering oleanders, takes in Rocklin where gangers once split the granite for the railroad; then cuts through Loomis, named after the Wells Fargo agent and saloon keeper: and so comes to Auburn (a fairer name for Rich Dry Diggings and Cold Dry Diggings) that leads to the Mother Lode, which with its golden network of veins, stretches from Mariposa northwards to the Feather River. The Spaniards, Mexicans and first Americans, however, had been interested only in the gold of cattle, crops and beaver alongside "El Rio de los Americanos", the very river which, for millions of years, had been carrying in suspension the new gold that it had been washing down from the quartz veins of the Sierras. All too soon the rivers were savaged by a new race

1 K. V. S. Alexander, 'Sutter Creek Cemetery', *The Detroit Society for Genealogical Research Magazine* (Vol. 26, No. 3, Spring 1963).

of men who came in their hundreds, and their forests of pine cut to the very ground to provide flumes to carry water to powerful hydraulic pumps or for timbering the shafts of mines.

A brand-new, brash and strident phenomenon called democracy was gestating in the beds of the rivers and in the dark tunnels below ground. The ubiquitous Yankee was biting the trees with his axe, tearing down the hills with his hydraulic hose and stabbing the ground with his pick. No longer did it seem important that King Charles the Third of Spain had once ordered an enquiry into the purity of the blood of any American woman who married a Spaniard, and that he held out no hope for the dignity of a civilized life once the Americanos swept into California[1]. Yet the ruthlessness of the mining camp was never quite able to destroy this older culture from southern Europe and so the Spanish names remained.

There are twenty-two Spanish missions and pueblos spread out within a day's horseback ride of each other along the 600 miles of dusty roads and tree-cooled alamedas from San Diego in the south to Sonoma in the north. But pride of place must fall to San Francisco; Yerba Buena, the beautiful city, to the Spaniards; the city of St. Francis to the Americans; and the Byzantium of the American West to the whole of the civilized world. It was growing slowly and naturally out of the sea and the hills until the explosion of the gold rush changed it overnight so that around the curve of the bay "hundreds of tents and houses appeared, scattered all over the heights and along the shore for more than a mile", and "every newcomer is overtaken with a sense of complete bewilderment"[2]. Many a Cornishman, homing to the Mother Lode from Wisconsin by way of the Mississippi and Nicaragua, shared in the bewilderment and then, like Arthur Lynne Sobey, a Cornish doctor engaged by a rich patient to be at hand while he was on buffalo hunts, came to love this eternal city. Some established themselves in business, like Josiah Phillips from Porthtowan who, after mining in Michigan and Nevada, invented the "Wee Pet" assaying machine for which he was awarded a gold medal in 1869 at the San Francisco Mechanics' Institute Fair[3]. Some walked the Embarcadero as surely as their own harbours in Cornwall, among them John Pearce of Newlyn who sailed in the *Balclutha*, one of the last remaining Horn ships and today a maritime museum.

But no career is stranger than that of Nathaniel Coulson. Born in

1 These letters are in the Museum of the Mission and University of Santa Clara.
2 Bayard Taylor, *Eldorado* (London: 1850), pp. 40-3.
3 *The Mining and Scientific Press*, 11 November 1871.

Penzance in 1853, he was left an orphan when his mother died before he was a year old and his father deserted him before he was seven. He lived for four years in the dreaded Workhouse, first at Plymouth and then at Bodmin, and at the age of ten started work as an apprentice on Penquite Farm, near Lostwithiel, that belonged to a Thomas Hoar. At fourteen he enlisted in the Royal Navy for five years and then emigrated to Scranton in Pennsylvania. In 1875 he joined a cattle train for California, worked as a handyman in San Francisco, sailed to New Zealand to find his father who promptly disowned him, and then found employment on a missionary steamer, eventually returning to Penzance where he gave public lectures on his adventures. In 1877 he was back in San Francisco again, clerking and newspaper reporting, and on the strength of his carefully hoarded savings made a bold decision. In 1880 he enrolled as a student in the dental school of the University of California, and qualified five years later, in time building up a lucrative practice in San Francisco which he then turned to good account for the benefit of Cornwall. He donated 500 dollars for the laying out of a public park in Lostwithiel, which still bears his name; he started an emigration fund to assist the young men of Lostwithiel and Bodmin; and he made substantial gifts to charities in the town of his birth. Though he lost much of his fortune in the earthquake of 1906 he was by no means ruined for, when the decision was made to build the new Grace Cathedral, he offered to finance the north tower and a carillon of bells since, like other Englishmen visiting the bay, he missed the sound of church bells and the art of change-ringing that is common all over England. The north tower was completed in 1937 and the bells were hung in 1941. On the great bourdon bell are inscribed the words: "First installed in Great Tower of Golden Gate International Exposition, February 18, 1939. Gift of Nathaniel Thomas Coulson. Born at Penzance, Cornwall, England, August 8 1853. Graduate of University of California, 1885". His munificence amounted to 190,000 dollars and no Cornishman has a finer memorial, for every day his name swings on wave upon wave of sound that creates anew the music that was first brought to California by the Spaniards and which Coulson himself first heard from the tower of St. Mary's Church in Penzance close by the terrace of granite houses that still carries his name[1].

But to most of the Cousin Jacks, in their hurry to seal their claims among the northern mines above Emigrant Gap, a bell was no more than a reminder of the lateness of the hour.

•   •   •

[1]   See Rosa Lee Baldwin, *The Bells Shall Ring* (San Francisco: 1940).

GRASS VALLEY, NEVADA CITY AND THE NORTHERN MINES

WITHOUT a doubt Grass Valley and Nevada City are the Cornish capitals of California for, ever since 1850 when George McKnight stubbed his toe against a chunk of gold-bearing quartz, or so the story goes, Cousin Jacks have been taking gold out of the rolling foothills that are dominated by their mammoth mine, the Empire Star. This was only silenced when, in the 1930's, the United States Government pegged the price of gold at 35 dollars a fine ounce. Today the two towns still hum with activity and glitter with prosperity for they have found the modern substitutes for gold: lumber and tourism. Grass Valley, at an elevation of 2,450 ft, is ideally qualified to hold the tourist title of the "gateway to the Sierras"; it is the envied possessor of four distinct climatic seasons with average temperatures of 75°F in the summer and 60°F in the winter; and in the fall it parades a riot of colours, the brilliant reds and golds of the planes, the oaks, the syca-mores, the maples and the poplars mingling with the dark greens of the silver-tip firs and the digger pines. Hoardings everywhere tempt the buyer to invest in Shetland ponies, trout farms, ranches, saw mills and gravel plants but everywhere too is evidence of a romantic past still living in the present; men and girls on horseback, equipped with lariots, colt revolvers and rifles driving herds of robust Hereford cattle; covered sidewalks; the Miners' Hospital; the Gold Nugget Inn and the Gold Nugget Cinema; the Bret Harte Hotel; and *The Nugget* newspaper that still circulates around diggings with the surprising names of Red Dog, You Bet, Town Talk, Rough 'n Ready, Humbug, Gouge Eye, Brandy Flat and Delirium Tremens.

The mines today are no more than historical monuments to this golden past and a rich quarry for the industrial archaeologist. The hoist of the Empire Star stands dramatic and spectacular as it pierces the ground at an angle of 27° and penetrates to a length of no less than 11,000 ft and a vertical depth of over 5,000 ft. Beneath its massive stamps, now broken and shattered, lies a vast network of tunnels drifted by Cousin Jacks for almost a hundred years, from which they dragged out more than 2,000,000 dollars in gold. If the Pennsylvania and the North Star, which they dropped one and a quarter miles below the collar of its shaft, are included, they accounted for the stupendous total of more than 500 miles of tunnel. Further afield, in a patch of grass and thicket at the Sneath and Clay mine, rusts a Cornish beam engine; and near Gold Flat, now a rich pasture for cattle but once a complex of boarding houses for bachelors, the mighty Idaho-Maryland mine surveys its own greying remains. The narrow-gauge railway that curved from Nevada City over high, slim trestle bridges, bringing the Cousin Jacks and their families from, and taking gold to, the main line depot at Colfax, has gone; and its locomotives have departed to Honolulu. Soon the museum of the Nevada County Historical Society in the old Fire House at Nevada City will be the only guide to this astonishing past. Already long toms and models of Cornish stamps are exhibited alongside Shoshone arrows and a glowing golden altar from a Chinese joss house, reminders that as well as the Cornish there were some 2,000 Indians in the Valley and an uncounted host of Chinese who provided the unskilled labour at the mines and on the railroads[1].

Here Cornwall is never far away. Fred Chegwidden, born in Virginia City of a father from Gwennap, walks out of Pengelley's shoe shop as he might do in Penzance. In 1957 the Mayor of Grass Valley was Richard Heather from Hayle. The President of *The Daily Union* newspaper is Robert Ingram from St. Ives and its managing director is Earl Caddy from Penzance. In the poplar-lined cemetery at the top of the hill, close by the Chinese burial ground, are the headstones to the Rowes, the Thomases, the Harrises, the Crazes and the Pascoes. In the high wooden Methodist church at Nevada City, founded in 1856, are stained glass windows in memory of the Clemos and the Curnows; and in the Episcopal church at Grass Valley, founded in 1855 and the second oldest in California, a memorial window to Mary Tredinnick.

Almost every resident seems to have some link with Cornwall. John and Fred Nettel (both of whom have died since this chapter was begun) remembered their father telling them how at the age of 12, he worked in the mines at Redruth by day and studied mathematics at night. In 1881 he emigrated to Michigan, then tried his fortunes at Prescott in Arizona and wild Tuscarora in Nevada before settling down in Grass Valley where a married sister was already living. When he had accumulated enough capital and had been promoted foreman at the Slate Ledge mine, following the normal Cornish custom he wrote to his girl in Cornwall asking her to marry him in Grass Valley. The first house they built was on squatters' land at Forest Springs near the mining camp of Allison's Ranch, and the second on twenty acres of free land, where they raised their family, both sons naturally going into the mines. John worked for a time underground at the North Star and

---

1 Most of the Indians were Diggers, living off pine nuts and grasshoppers.

then for fourteen years in its mill, all the time measuring up to his local responsibilities; he was president of the Nevada County Historical Society, and for twenty years under-sheriff, in itself a token of the trust that the electors placed in him. His brother, like his father, worked and studied at the same time, assaying at the Empire Star and learning about metallurgy through correspondence courses. Considering then whether to go to Chile, for the blood of the migrant miner ran strongly in his veins, he happily exchanged this idea for marriage and settled down to the double occupation of assayer and accountant to the North Star Mining Company and the New England Consolidated Mines, located on Gold Flat. Their friend John P. Mitchell, too, has passed away since this chapter was commenced. In 1960 he was one of the oldest surviving Cousin Jacks and a dying reminder of the hazards of his occupation, for his lungs were choked by a lifetime's accumulation of grit and dust and he could only talk in laboured whispers. Known as "John P." to distinguish him from "John Q." and "John R.", for at one time there were so many John Mitchells in Grass Valley that a further alphabetical identification became necessary, he hailed from Illogan and for a time had worked at Tincroft, one of the oldest mines in Cornwall, but in 1896 unemployment finally forced him to emigrate to Johannesburg. He then returned to Cornwall, set off again to Hancock in Michigan and finally found his way back to the First Wesley Church at Tuckingmill to marry. The last flight to security was to Grass Valley, though he had no job awaiting him, but as his wife simply explained, "I had an uncle there and I knew the Thomases and Ben Opie". So he successfully rose from the position of miner to shift boss at the Pennsylvania and the Empire Star and, when he was too old for active service, became a watchman and specimen boss. "High-grading" or helping himself now and again to a piece of unusually valuable ore was a miner's occasional lapse from virtue in Californian mines, against which the owners had to protect themselves through their specimen bosses whose duty it was to search suspects. One Cousin Jack was so clever at telling in the dark whether a chunk of rock was worth secreting in his pocket that he was nicknamed "Old Velvet Thumb"; and there was another who always complained of stomach pains at the time of blasting and had to be sent home, until one day his dinner bucket was examined by the specimen boss and the truth revealed[1].

Cousin Jack tales of this kind meander like the veins themselves and

---

[1] High-grading is not to be confused with what the Cornish termed "family ore", the rich golden rock they reserved for themselves.

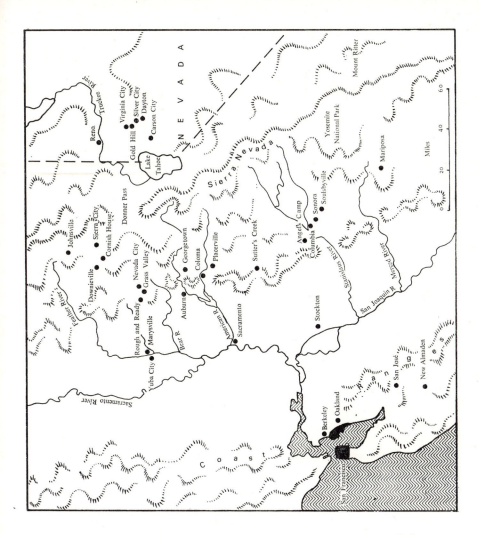

*Cornish settlements in
California and Nevada*

none is more skilled in the telling than Ray Polk; his real name is Polkinghorne but his father considered it too long for a signature. His grandfather, "a hard drinker and a good fighter", left Redruth in 1866 for Grass Valley by way of Panama and there married the daughter of Nicholas Morcam of Redruth. Today their grandson with his unique collection of bells recalls nostalgically the Christmas Eves of other years when, with snow on the ground and the stars frozen in the sky, he could see the miners returning from the shifts at the Champion, the North Star and the Empire Star, making their way down the hillside, swinging their lamps and singing their Cornish carols. Today the canyons still ring with their melodies when the Grass Valley Cornish Carol Singers assemble on the steps of *The Daily Union* newspaper office. Indeed their reputation extends throughout the whole continent and beyond, for they have sung from the 2,000 ft level of the Idaho-Maryland mine in a nation-wide radio hook-up, from the Capitol building in Sacramento and in the streets of Virginia City in Nevada, as well as making recordings that have gone to Cornishmen the world over. Every January they present a service of carols in Oakland at the Shattuck Avenue Methodist Church where the pastor is a Peters, the custodian a Curnow and the worship leader Will Ralph whose mining ancestors lie in St. Hilary. The service starts with the traditional "Roll-call of Cousin Jacks and Jinnies"; and round the spacious well of the church echo the names of Cornish towns and villages, memories of which have not been forgotten although the pronunciation of them has changed with the misting of the years: Truro, Ponsanooth, Green-bottom, Tywardreath, Lelant. Their president is Jim Hollow from St. Just and their musical director is Harold George from St. Austell, whose wife was born on St. Michael's Mount of a father who worked on the Californian railroads and drove the first steam train in Mexico. George was a "primitive" in music until the choir trustees despatched him to San Francisco State College to study for his "credentials"; he then kept a music shop and taught at the Grass Valley High School, where his son now teaches and helps with the training of the Cornish choir. Together they carry their music into the fine new Methodist Church that has been built in the style of a Spanish mission with white walls and a red tiled roof, its organ a gift of Simon Craze and its many coloured windows in memory of the Tamblyns, the Jameses, the Bennetts, the Edwardses, the Argalls, Mrs. T. Prisk, William J. Bray, J. H. Carthew and the parents of Harold George.

The distinction of being the first Cornish miner in California may be of no consequence, but a claim has been made on behalf of Captain

James Rickard, the father of a famous mining family from Porthtowan. One of his grandsons, T. A. Rickard, equally distinguished in American mining circles, states in his *Retrospect, an Autiobiography,* published in 1937: "He was one of the first accredited miners to go to California in those golden days: he crossed the Panama Isthmus and took with him a sectional 10-stamp mill, the first to be put to work in California, in June 1850, at Coulterville". It appears that he had been engaged as a superintendent for the Nouveau Monde Mining Company that had taken over the assets of the Merced Mining Company. There were at least twenty other Anglo-Californian companies registered in London, with a total capital of 10,000,000 dollars, all engaged in sending out agents to negotiate for prospects and veins; and one of them was the Anglo-Californian Goldmining and Dredging Company which in 1850 despatched their director, Sir Henry Vere Huntley, to inspect their titles to claims on the Stanislaus and Calaveras Rivers. He apparently took with him nine Cornish miners and put them to work with the Mariposa Mining Company that was operating only twenty miles from Coulterville where Rickard was, but who they were is not known[1]. It is more than likely, however, that there were even earlier arrivals, miners who were prospecting relatively nearer to California when the news of the first gold strike broke. For instance, the writer and painter, J. D. Borthwick, says that in 1852 at Placerville (Hangtown) he met several Cornish miners who told him they had come "from mines in Mexico and South America and from the lead mines in Wisconsin"[2]. That this was not merely hearsay is supported by evidence from Cornwall: the 1825 annual report of the Royal Geological Society of Cornwall mentions "the great number of our countrymen who have gone out to America on the new mining enterprise"; and the minutes of 1833 and 1834 note Richard Pearce superintending a silver mine at Arica in Peru and John Rule the Real del Monte mine in Mexico. That they were also rushing in from Wisconsin is equally certain; many of the obituary notices of Cousin Jacks killed in Grass Valley refer to their families in Wisconsin; one of the first hotels built in Grass Valley was called the Wisconsin, which suggests a connection with home they meant to preserve; and Louis Copeland, who interviewed many old-timers, reckoned that Wisconsin was almost denuded of the Cornish until many of them returned disillusioned. And one who died in California

---

1 Norman Harries, *Cornish and Welsh Mining Settlement in California* (unpublished M.A. thesis, University of California at Berkeley, 1957).
2 J. D. Borthwick, *Three Years in the Goldfields* (London: 1857).

shortly after his arrival, perhaps out of sheer exhaustion, was John Hay of Yorkville, a member of the community of Methodists there [1].

Another was Edward Skewis of Camborne who perhaps was among the first of the Wisconsin pioneers to follow the ruts of the wagon wheels of Captain Townshend and his Rough and Ready party who arrived in California in September 1849; but of him we know rather less than of his brother James, who joined him two years later, for at the age of 92 he dictated an account of his adventures. Preferring the Mississippi route, and accompanied by another brother, he travelled by stage coach from Shullsburg to Galena, and at Peoria on the Illinois River boarded a steamer for St. Louis and New Orleans. About the delights and hazards of the down-river voyage or of the perils of the sea passage to Havana and thence to the Isthmus of Panama he is silent, but of his entry into California he has left a very animated account:

In a few days we landed in San Francisco at that time a very small place, but there was much business and gambling going on and everything very expensive. Then there was a boat ride to Sacramento for 8 dollars, then by stage to meet Edward at Deer Creek about four miles from Nevada City, where we opened up a rich claim for the summer's work and did very well. Most every day between three of us got six ounces of gold until the rainy season set in. Then we washed in smaller creeks between the hills, which also paid satisfactory. The gold was much larger in size, what was called rough or nugget gold. Just then there was a strike at Grass Valley on a slide from a hill and we got a claim and sold it for 100 dollars. One day a man gave me ten dollars for one day's work on that hill to do some timbering work as he was not used to that kind of work, and that was the wages paid at that time. Then we went further south to Sonora and Springfield. However we did not find it any better than the place we left nor as good for some days we made 15 dollars in gold dust.

That was the end of their three years of prospecting along the creeks of the Mother Lode for at Sonora they heard of a gold strike in Australia and so took the first ship out of San Francisco. But they were disappointed and after a year sailed in the *Anglesia* for England; they actually persuaded the captain to transfer them to a fishing boat bound for Penzance as soon as the cliffs of Cornwall were sighted. Camborne was only twenty miles away but it was their home no more and they were soon back at Shullsburg. Having circumnavigated the globe before he was thirty in search of his crock of gold, James Skewis tried to settle down to a life of farming but with no success. The Civil War approaching, and owing allegiance to neither North nor South, he disappeared "underground", sailing with his family to Mexico from

---

[1] From the *Minutes of the 1851 Annual General Meeting of the Board of Trustees of Yorkville Church and Cemetery, Wisconsin.*

New Orleans "on the last boat before the harbour was closed on account of the war".

Another Cornishman from Wisconsin searching the creeks of the Mother Lode at the same time as Skewis was Charles Strongman who had emigrated to Mineral Point in 1841 from Perranzabuloe. In letters to an acquaintance in Wisconsin[1] he describes the five-month journey overland in the spring and summer of 1852: the trail through Iowa to Council Bluffs was "over the worst kind of roads"; and Salt Lake City was "not much of a place". In California he met many Cornishmen from Wisconsin who had followed the river route: "they had a very hard time of it, paying four or five hundred dollars to be starved to death". He himself was not impressed with the opportunities in California for, while he could earn three dollars a day with board, the cost of living was far higher than in Wisconsin. Significantly he adds that there is no need to describe conditions at the diggings, for many disillusioned miners had already returned to Wisconsin to warn others to remain where they were. He too was on the point of making the return journey overland in 1858 but was robbed and murdered in Placerville.

Whichever route the immigrants were forced to choose to reach Grass Valley, each had its problems and terrors. For the Cornish in Wisconsin the Mississippi run was longer and more costly but the only real danger was the passage across the Isthmus of Panama which might consume five days by mule over mountains and in canoes over swamps with the risks of yellow fever, though the experiences of Bayard Taylor were more reassuring: "in spite of the many dolorous accounts . . . there is nothing, at the worst season, to deter any from the journey"[2]. Those coming direct from Cornwall favoured it for the obvious reason that they had the sea with them nearly all the way; few were rich enough to attempt the longer 17,000 mile stretch round the Horn. The first settlers in Grass Valley, however, appear to have been the 1850 immigrants who, following the Truckee River route, were searching for their famished cattle that had wandered off the main trail[3]; they had no idea of the fortune they had stumbled on until quartz veins were discovered on what they called Gold Hill and Massachusetts Hill. They staked their claims and erected make-shift mills but could make little progress since "there were no miners in the country who knew how

---

1 These letters were discovered by Norman Harries.
2 Taylor, p. 20.
3 Brown and Dallison, *Directory of Nevada, Grass Valley and Rough and Ready* (1856).

to open or work a quartz ledge"[1]. And as it happened hard-rock mining here was to present special difficulties for the veins were very narrow and the strikes and dips very variable. The Aqua Fria, an English quartz company that had been floated to develop claims at Mariposa, seems to have been the first to tackle these problems, for an examination of its accounts reveals that considerable sums of money were spent on transferring equipment and skilled miners from England. It seems reasonable to suppose that among them were Cornishmen, and that these may have been moved from Mariposa to Grass Valley when the company bought the Gold Hill mine about 1854. They may be the Cornishmen who are listed in Brown and Dallison's *Directory of Nevada, Grass Valley and Rough and Ready* of 1856 as living in either cabins or boarding at the Metropolitan and Oriental hotels: John Bluett, A. K. George, John D. Jenkins, Lemuel Nye, William Perrin, Thomas Panglase, together with two Rowses, two Treloars and three Mitchells. They appear to have been justified in their selection for they are credited with erecting in 1855 at Gold Hill the first Cornish pump in California, buying the mine from their employers and working it "with little interruption for eight years longer"[2].

It seems strange, however, that no other Cornish are mentioned, for in 1856 Nevada County boasted a population of 25,000 and Grass Valley was being described as a settlement of "mixed primitiveness and business bustle" that only awaited "the pleasant and harmonizing influences of female society". There is no trace at all of the Cornish contingent from Wisconsin so a possible explanation is that they were migratory, like Skewis, moving among the diggings along the great bend of the Yuba River at Washington and Jefferson and later going to Newton, Kentucky Flat, Eureka, Snow Point, French Corrall, Sweetlands, Cherokee or Little York. For these early arrivals life had been exacting, the first winter of 1849-50 being most severe and the transport of supplies from Sacramento difficult. The snow lay 10 ft deep and the cost of living soared; a new pair of boots could not be bought for less than 40 dollars or a long-handled shovel for less than 16 dollars. Consequently in the fall of the following year, when it was reckoned that there were 6,000 miners in Grass Valley, vast stores were ordered in anticipation of an equally severe winter. But it proved to be mild and rainless so business speculators became bankrupt and left. Since a shortage of rain meant there was little water to work the rockers and long toms, the miners left too.

[1] E. F. Bean, *History and Directory of Nevada County* (1867), p. 49.
[2] Ibid., p. 55.

For similar reasons it is impossible to give precise figures about the Cornish ten years later. The 1860 census returns for Grass Valley are not helpful; out of a population of 3,940 the English account for 530, of whom 470 were miners, but we do not know how many of them were Cornish; and in any case many had already left the valley because of a depression. E. F. Bean in his *History and Directory of Nevada County* of 1867 reports that between 1859 and 1861 at least one third of the male population left for the Comstock, while the Cornishman John Cord in a letter of January 1858 to his brother at Mineral Point says: "Grass Valley is a poor place at present . . . money is scarce . . . I am not making anything . . . I live on interest now . . . A man with 40 acres of land in Wisconsin making grub is better than he can do here . . . many families wish themselves back again"[1]. And two years later he found himself beaten by the depression and returned to Mineral Point rather than face the rigours of the Comstock. In November 1861 it is true that *The Daily Union* was asserting that "our mining is for the most part carried on by Cornishmen", but Hugh B. Thompson's *Directory of the City of Nevada and Grass Valley* for the same year lists no more than sixteen who might be regarded as of Cornish origin. At Nevada City there were two miners, J. Eddy and T. Stevens, and one blacksmith, S. T. Oats; at Grass Valley the miners were three Bennetts, Anthony James, William Mitchell and Thomas Pielglase (Polglase?). Henry Rule was a machinist, Edward Rule a blacksmith, and Edwin Uren and William Bettis carpenters. Even if Anthony James, the president of the Empire Quartz Mill Company, his secretary Charles Jenkins and the solitary figure of Richard Bullock at the city brewery are included, the number of the Cornish remains small. Two of them, William Mitchell and Thomas Polglase, might be considered as employees of the old Aqua Fria enterprise, but the remainder are unaccounted for. If the 1860 census return of 530 English included a fair proportion of Cornish, then it seems probable that many had left the valley either for home in Wisconsin or to the Comstock, driven away by a seasonal depression or the flooding of the mines in the winter of 1861-2, when one unfortunate man, Peters, was trapped and drowned in forty feet of water in the breast of the York Hill mine[2].

By 1865, however, as Grass Valley enjoyed its war-time boom, the population jumping from 4,000 to 10,000, the picture of the Cornish is very different. With its 19 quartz mills running 208 stamps and its 38

---

1   Harries.
2   Florence M. Rittenhouse, 'The History and Romance of Grass Valley', *The Grass Valley Daily Union*, 20 January 1962.

mines employing 1,500 miners (there were another 1,000 prospecting on their own), William S. Bryne in his 1865 *Directory of Grass Valley* describes it as "the wealthiest, most prosperous and most extensively known of the interior towns of California". Moreover he records the names of about 200 Cornish.

Some commanded positions of responsibility; Thomas Crase was superintendent of the Gold Hill mill and town marshal; Mark Bennallack was underground superintendent at the Eureka and Josiah Goldsworthy its foreman; William Rodda and William Hoskin held similar posts at the North Star; and William Colliver was foreman at Hues Hill. Some were even part owners of mines, like John Trenberth and fifteen Cousin Jacks who bought Hueston Hill. Trenberth seems to have been a shrewd and successful business man for he was able to dispose of his $1/16$ share to William Rodda for no less than 11,500 dollars, a fortune never to be dreamed of in Cornwall. He was also one of the three owners of Town Talk, which in 1865 was taking out some 5,000 dollars of gold a month; he owned shares in Scadden and Company who worked the veins on Boston Flat and in the Stockbridge Company mining to the east of the claims of the Rocky Bar Company on Massachusetts Hill; and the Betsy mine was also his, bought for 9,000 dollars, to which he gave the Cornish name of Wheal Betsy. The fabulous North Star in 1865 was also under Cornish ownership, having been purchased for 15,000 dollars by J. C. Pascoe, William Kitto, William Hoskin and the three Rodda brothers, William, Josiah and John, the latter "an excellent citizen and universally esteemed", who combined mining with horticulture and found time to drill troopers as a Lieutenant in the Grass Valley Union Guard.

By 1865 then the Cornish were formidably entrenched in the goldfields, enjoying the prosperity of the war years with "quartz on the brain, fits of ambition, fits of envy and fits of extravagance", as *The Daily Union* described them. Though in the general air of affluence it was disagreeably discovered that the mortality rate among children was far too high and a Sanitary Fund was started with William Rodda donating 20 dollars, there was enough surplus money around for the building of a new theatre in Nevada City, for engaging visiting lecturers on such subjects as "the Education of the Industrial Classes", and for arranging an exhibition round the theme of Milton's *Paradise Lost*. Methodists were urged to visit it on Sundays, when they could meet their God "face to face", and infidels the remainder of the week "to see Eve before the Fountain". And for those of musical tastes the pianist Gottschalk gave recitals by night and tuned their cottage

Cross-section of a lead mine in Wisconsin in the 1840's. The equipment needed by these Cornish "badgers" was extremely simple. [*State Historical Society of Wisconsin*]

Emigration broadsheet issued in Padstow in 1844. This brig was probably engaged in the American timber trade and carried emigrants on the return voyages as "paying ballast". [*Royal Institution of Cornwall, Truro*]

pianos by day. But all too often full pockets were emptied in drunken brawls, usually early on Sunday mornings: "A number of Cornishmen amused themselves and the whole neighbourhood in the vicinity of Richardson Street by indulging in a free fight near the Brewery". On one of these occasions J. R. Rowe and Edwin Drew were arrested by Constable Meek for fighting in the cellar under Gad and Company's clothing store: the judge fined Drew 16.50 dollars and Rowe twice the amount because he refused to plead guilty. So arduous was the constable's work that three new cells had to be constructed in the county jail. But in a mining community like Grass Valley tragedy was never far away from hilarity. Mrs. Crase fell headlong to her death in a hidden shaft near the Gold Hill mill; and a snowslide buried George Nichols. When the spring sunshine and the thaw uncovered his body it was in a perfect state of preservation, but by that time his demented wife had already left and there was no relative to pay the last respects.

The boom was to continue for another five years. Buildings, mines and families proliferated. Every week came news of a fresh venture; of a rich ledge of quartz discovered in the bed of Deer Creek below the old Cornish Mill; of the need for new crushing mills at Rough 'n Ready; of a new strike at Yellow Jacket; and of a gold brick, weighing 241 ounces, taken from the Lucky mine, the result of only one week's run. According to Bean's *Directory of Grass Valley* for 1867 as many as 500 new houses had been built in the previous year and a half to accommodate the new immigrants, for it was in the last three years of the 1860's that the Cornish began to arrive in their hundreds. One reason was that the depression was at its worst in Cornwall but another was that the Panama route had been cut by several days through the completion in 1867 of the Central Pacific Railroad from Oakland to Cisco along the southern border of Nevada County. Though the immediate purpose of the new track was to serve the silver mines at Washoe, it had the advantage of bringing the Grass Valley mines into closer touch with men and capital from Europe, while San Francisco, as a centre of supplies for the miners, was now only a day away. The result was that at Grass Valley "the very best Cornish miners could be had for 3.50 dollars a day"[1]; the ore dressers were all using Cornish methods[2]; and, according to Bean, 250 Cornish "nomads of pick and shovel" had settled their families in Grass Valley and Nevada City.

---

[1] *Facts and Statistics relating to the Edmonton Gold Mine, Grass Valley etc.* (Boston: 1866), p. 16.
[2] B. Silliman, *Notes on the Quartz Mines of the Grass Valley District* (Nevada City: 1867), p. 9.

Yet, in spite of the high level of prosperity, the Cornish had their personal problems, not least those of bearing children and keeping them alive. *The Daily Union* complained that during the winter of 1869-70 the hearse had been seen in the streets far too often, carrying off the children who had succumbed to scarlet fever. In 1870, out of a population of 7,000, death from "natural causes" accounted for 123, almost 2%, equivalent to a funeral every three days; but only three died from injuries in the mines. No one, however, thought of returning to Cornwall; and some, like George Richards, Richard Pascoe, William Trevithick, William Betallack and John Rodda had already sealed their future by becoming full American citizens.

At the North Star, a veteran of the Mother Lode mines, the underground superintendent, James Bennallack, had installed new pumps to prolong its life, and with his brother had uncovered the South Star near the head of Spanish Creek. Together they had sunk a shaft to a depth of 160 ft, run a drift of 300 ft at the 100 ft level, and were introducing hoisting machinery worked by a boiler that also served to heat a drying room. Nevada City daily resounded to the clatter of 480 stamps, crushing 250,000 tons of ore a year. In March 1871, in the Cornish mine on Deer Creek, pay was found rich in sulphurets of iron, copper and galena. Owned by William and Philip Richards, Mrs. Elizabeth Richards and Capt. Sam Adams, in three years it had earned them 30,000 dollars, more than enough to send Mrs. Richards on a vacation to Cornwall where she no doubt told her relatives of the boundless opportunities that still awaited them in California. Perhaps she recounted the story of the sensational chance find by one Reese who, when sluicing off some old ground on Randolph Flat, picked up a nugget worth 500 dollars. So prosperous was the Cousin Jack that on 16 April 1870 *The Daily Transcript* observed: "A Cornish working man who has resided here for nearly twenty years says from what he knows of his countrymen in this county that they have thousands of dollars in coin buried for the purpose of escaping taxation". *The Daily Union* of 28 May 1871 was somewhat kinder, reprinting the following appreciation taken from *The Owyhee Avalanche* of Boise, Idaho:

The Cornishman is probably the most skilful foreign miner that comes to our shores. For this he deserves no special credit, because it is a calling to what he has been accustomed since his childhood. The Cornish miner is of quiet disposition, although very headstrong, probably on account of his being a Johnny Bull. Generally speaking, he is satisfied to be working for others, but insists on being paid promptly for his services, and doesn't care for engagement in mining on his own account. They are mostly stalwart, good-looking fellows, dress better than any other class of miners and are very fond of women. They also appear more clannish than any other foreigners and a majority of them are very good singers. When they visit a saloon they generally arrange themselves around a table, call for a pot of ale or porter, and pass the time away in anecdote and song.

But the same paper also published some verses to remind its readers of the dangers of his trade:

> 'Tis his to find the glittering ore
> For ages hid in earth's dark womb,
> To creep and climb, to dig and bore
> And build himself a living tomb
> Some six feet high, some four feet wide
> And reached o'er depths that few would stride.

The reminder was deliberate and timely for 1871 was a year of terrifying accidents. Joseph Treweek, living at Boston Ravine with his wife and three children, was mutilated by a runaway truck at Allison's Ranch; James Peters was killed outright by a cave-in at the Empire Star, leaving a wife to mourn him in Cornwall; and William George died instantly when he fell to the bottom of the Good Friday shaft of the Ophir; they were all under thirty. Richard Stevens was walking round the shaft entrance of the Yellow Jacket, holding his candle before him, about to fetch a drink of water when, for some unaccountable reason, he walked directly into the mouth of the shaft and plunged 300 ft to his death. William Gribble, who had left his wife and family behind in Wisconsin, had his face smashed to pulp by falling rock. Joseph Hoskins and Dick Richards were severely injured at the mouth of the tunnel of the Blue Gravel Company at Marysville when a box of cartridges exploded in their faces. John Kinsman and William Craze were blasting in a stope at the 500 ft level of the Lane and Fuller shaft in Nevada City when there was a premature explosion. Kinsman's hands were blown to pieces and Craze's skull shattered to fragments. The merely injured were only relatively fortunate; John Gribble in the Idaho had his right arm crushed to the bone by falling rock and an artery severed. Henry Odgers in the Empire escaped death with a fractured leg; and George Pascoe lived to work again after bolting horses had raced an ore wagon over him. For their dependents, of course, incapacity or death meant the suspension or end of all income, unless the miner had insured himself, and this was not the general practice. So when John Bawden was killed at French Lead and it was discovered that he had insured himself for 5,000 dollars, *The Daily*

Union did not hesitate to drive home the lesson that all miners should follow his example.

Actually a Cousin Jack, if he were skilled, could well afford to be prudent for the sake of his family. Though three dollars a day was considered a low wage compared with four on the Comstock, yet it was equivalent to 900 dollars a year, which was more than adequate if he had built his own house and rented half an acre of land on which to grow his vegetables.

There was always periodic grumbling about low wages in Grass Valley, but usually it was the Irish who complained on the grounds that "class mining" operated a nationality bar against them. The truth was, however, that wages tended to be depressed because of the Chinese. In 1869 Chinese labourers at the Empire mine had been roughly handled by both Cornish and Irish alike; and the sheriff had been compelled to swear in a posse to protect property. Then again in July 1871 troops had been used to break a strike at Amador City, whereupon the Cousin Jacks in Grass Valley banded themselves into the Working Men's Society and passed a resolution that "all those of Mongolian origin should be prohibited from entering California". It was not that they wanted to indulge in recriminations about race so much as to maintain standards in their industry; nor would they permit themselves to become the instruments of political agitators, for when James Phelan, the president of the Miners' Union of Virginia City, visited Grass Valley, they nearly ran him out of town.

Few of the Cousin Jacks were union-minded for they knew that their skill, experience and reliability could always command a high price; and if feelings ran too high there was always some activity to siphon them away, and not always the saloon. They had helped to found the Lyceum where William George might read his essay on "Home and Its Influence"; where they might debate compulsory education; or where a Professor Holmes might lecture on Charles Dickens. Not all of them, of course, took kindly to this "culture", as *The Daily Union* observed somewhat tersely: "Grass Valley is not considered a good town for lecturers, many persons here preferring to play jigger or seven-up to listening to a good talk". Yet Grass Valley boasted a Literary Association with a library of over 700 books, and claimed in 1871 that 407 had been loaned mainly on history, biography, travel, science, theology and "the highest order of poetry". If it is too much to suppose that a Cornish élite was emerging, surrounded by its halo of sweetness and light, yet they were not totally indifferent to education; and they were among the first to criticise the high tuition

70

fees at the new State University at Berkeley, complaining that it was the "rich man's university, built at the expense of the many".

A more spectacular institution for canalising surplus energy was the annual wrestling tournament, held on 4 July, and open only to Cornishmen. It took place in a ring, 60 ft square and made of boards, which was covered with a light canvas to give the appearance of a tent and so contrived that it admitted plenty of air and excluded the strong sunlight. On all sides of the ring were seats for 800 spectators and in the middle the arena, which was covered with straw, where the contestants staged their bouts. These trials of strength were decided beforehand by a committee, but the actual contests were judged by three "sticklers" who, while a "play" was in progress, stood around the wrestlers and watched every movement. A fall was considered to have occurred if two shoulders and one "pin" were on the ground, the shoulder being defined as the shoulder-blade and a "pin" as one hip joint. The wrestlers fought in bare feet and were clothed from the waist down in light pantaloons and thin drawers, while above the waist they wore a loose jacket of strong canvas fastened in front with cords. Holding each other with the "grip of grizzly bears", they would fight until one was thrown, the victor going on to fight the winner in another contest. Throughout a hot afternoon and a sultry night as many as forty bouts might have to be contested before a final decision was reached, when the champion would be rewarded with a purse of a 100 dollars in gold.

By 1872 the seven fat years had ended and a depression was already in sight. Some mines, like the Idaho, were still operating at a profit, though a reduced one; for instance, in one month's run its fifteen stamps brought in gold to the value of 43,000 dollars; its payroll for the same period was 14,000 dollars; and there was talk of re-equipping the mills at a cost of 75,000 dollars. Likewise the Empire on Ophir Hill, though its stamps were running to full capacity night and day, was feeling the need to replace its worn-out machinery. The reports of other mines were even less assuring; the North Star's pumps had been damaged and the workings flooded; the Cedar Quartz on Wolf Creek had almost stopped because of haulage problems; and the Greenhorn mill was idle through lack of water. Howard Hill had been silent for many years, and in the winter of 1872-3 ten others shut down or struggled on short time. It was difficult to convince the Cousin Jacks that they were unemployed because of the increasing costs of extracting the ore and the need to attract new capital into the industry; and they were not always willing to co-operate with management when new techniques were introduced to reduce costs. One of these was the introduction of Giant Powder, a far more powerful explosive than the old Black Powder, which, according to John Jewell of Silver City in Idaho in a published letter to C. H. Mitchell of Grass Valley, allowed three men with single-handed drilling to perform as much work as six with double-handed drilling. But the Cousin Jacks foolishly suspected that this improvement would mean further unemployment and so complained that the fumes made them sick. Although Jewell pointed out that the new explosive would in fact increase their wages since they would blast out more rock, for which they were paid by footage, and an analysis of the fumes at San Francisco proved they were not injurious, they remained unconvinced. Bickering and argument below and above ground sparked off strikes; at the Empire twenty-two miners downed tools and two others who refused to join them were shot at on their way home to their boarding house.

So, in the spring of 1873, the trek to the Comstock began over the precipitous heights of the Sierras by whatever transport was available, in wagons or on the regular and special stages. Among the Cousin Jacks was their champion wrestler, William Reynolds, of whom *The Daily Union* wryly remarked: "The boys on the Comstock had better not tackle him in a Cornish wrestle". Another was John Trezona from Camborne and Houghton in North Michigan. Leaving on one of the first overland trains from Chicago to San Francisco he found in Grass Valley a wife, whose maiden name was Rowe, but no work; so the pair joined the stampede to the Comstock and settled at Gold Hill, where a son was born, the father of Charles Trezona who is now director of Trezona-Schiller Associates Ltd. of Los Angeles. They were all sure that fortunes would tumble into their laps once they alighted on the Nevada side of the mountains in spite of *The Daily Union*'s persistent advice, "stay at home and build up the home place"; and a year before it had printed, as a warning, some verses by Mrs. Louisa Liddicoat about her brother George who, in Humboldt County of Nevada, had been shot dead by a stranger "with murderous heart so rife" as he was mounting his horse. Though the scene of the murder was many miles from the Comstock and near the Oregon border, yet Virginia City had an ominous ring for those who preferred more settled conditions for their families; these turned their horses south and entered another paradise at the quicksilver mine of New Almaden.

Grass Valley was hardly depopulated but the daily sight of so many empty houses made it a restless and disturbed town. For the first time a divorce was reported in the newspapers, "Mrs. Ennis not having acted as a wife should act". Then one stormy April night Henry

Roberts from Murchie's Lead sought shelter in the boarding house of Belle Butler; when he was sleeping on a chair in the parlour, there was a knock on the door and Belle Butler admitted a young girl, Kitty Hess who, on seeing Roberts, screamed, "you're a bloody Cornishman", and shot him dead with her derringer. Again, one Sunday morning, George Mitchell arrived at the boarding house of Matthew Rogers to find him brandishing a butcher's knife and in violent argument with Peter Daws who had armed himself with a miner's pick. To part them, Mitchell fired his pistol in the air but he was too late, for already Rogers was lying on the ground with the pick sticking into his back. Though a Grand Jury acquitted Daws because he had been provoked and acted in self-defence, the macabre episode sobered the entire town so that no liquor could again be sold in saloons on Sundays.

As if to strike a higher level in public behaviour, the Methodists now decided to build a new church. The first had been consecrated in 1852 but was too small, being "made of wide boards stood on end, battened, limed and papered, with one aisle from door to door", and had lasted only two years. The second, much wider and slightly longer, had been ravaged by dry rot. The third was to accommodate a congregation of 1000; three trustees supervised its construction, James Peters, William T. Rule and William George; and within nine months they had collected more than 4,000 out of the 7,000 dollars needed. Nor did they overlook the education of laymen and their wives. Professor George, the President of Napa College, travelled the 150 miles across the Sacramento valley to lecture them on "The Genesis of Creation or a Bird's Eye View of the Universe"; and Dr. Paul Brennan spoke on "Anatomy and Health", making sure of an audience by calling single lectures "The Throne of the Mind"; "The Path of Life"; "Love and Marriage" (for ladies only); "How to Live"; and "What to Live for".

Fortunately for those who had chosen to remain in Grass Valley and ride out the depression of the previous year it proved to be only temporary. With its 8 hotels, 4 breweries and 32 saloons the community could still boast that it was the "quartz crowned Empress of the Sierras", serviced by 5 cobblers, 4 bakers, 6 lawyers, 8 doctors, 2 undertakers, and 3 dentists who performed their extractions by electricity, an anaesthetic being given only "if desired". More replacements arrived direct from Cornwall now that the transcontinental railroad was completed and were soon found employment. For those like William Coombes, with a little cash to invest, a claim could be bought on Randolph Flat for 450 dollars with the expectation of a profit of 200%; and for those without capital, jobs awaited them at the Howard

Hill, near the Union Hotel, and at the Erie, where new shafts were being sunk. Further north, at Sierra Buttes, there was a call for fifty men though its altitude of 8,500 ft made some miners think twice about going on account of mountain sickness, shortness of breath and lassitude. Moreover the nearest doctor lived at Downieville and charged 35 dollars a visit; and "one man had the doctor four times in five days and had to die to save charges". But in Grass Valley there was to be no shortage of work for the next seven years for, as *The Daily Union* reported in 1880:

No dwellings are idle . . . happy and contented families are seen everywhere in their flower-embowered cottages and in the many elegant residences in the town. A considerable mortality among the workers would do the place some good we think, and our undertakers, who seem to be the only idle men in the town, will agree in this opinion. Gold everywhere and the richest and best fruits growing above the gold.

James Bennallack, for instance, was re-opening the old Sebastopol, which had been idle for many years, and was transferring to it the pumps and machinery from the deceased Bullion mine. Harry Thomas and Charles Mitchell uncovered a new mine which they called the Alpha as if to herald a new beginning for the community. Four other Cousin Jacks, Philip Trezise, W. H. Berryman, John Jenkin and William Tregonning were floating a new company, the Republic Mining Company; and an older one, the Centennial Gold Mining Company, was showing considerable profits under the Cornish management of Israel Quick its president, W. H. Mitchell its treasurer and landlord of the Wisconsin Hotel, and John Trembath its superintendent. At the rear of the Catholic church John Tierney was installing in the new Peabody mine machinery salvaged from the derelict Oakland mine. John E. Carter and John Glasson were sinking the first shafts of the Black Lead property on Massachusetts Hill. With sixteen saw-mills biting through 20 million feet of timber a year and seventeen quartz mills pounding 85,000 tons of ore, there seemed to be no end to prosperity.

But in actual fact by 1890 the production of gold was shrinking, though its effects were not yet apparent. The younger generation of Cornish, however, seemed restive and anxious to make their fortune in the latest of the camps in Colorado, especially Leadville, undeterred by the accounts in *The Daily Union* that it was a town where on Sundays "blear-eyed and bloated debauchees, in whose besotted lives honour, decency and manhood have long since died, bandy ribald jests with lost women who carry the liquor to the tables". An older generation found it difficult to convince them that it was far better to stay in Grass Valley and sip the pleasures of a sophisticated society that had been achieved by years of effort; to participate in harmless walking matches; to watch William Hocking in a drill-sharpening contest open to competitors from all over North America; to cheer Jim Rodda to victory as he boxed Thomas McAlpine, the middleweight champion of California, for a purse of 300 dollars; or even to sing oratorios with the Handel and Haydn Society. There was also the attraction of the Grass Valley Brass Band, founded about 1875 by John Thomas from Porthtowan, in which they could join the elect with Richard and John Angove, Richard Goldsworthy, Richard Jewell, Thomas Tredrea, Fred Rowe, and Charles Pascoe. For intellectuals there was the exclusive society known as "The Ten" which met on Monday afternoons to study English literature. For the more practical the leisure hours could be filled with gardening or building a pond and stocking it with carp bred by John Rodda. Who, it was argued, would want to exchange the peace of the Cornish sheriff, James W. Peters, for a Leadville of 40,000 miners where it was possible for an ex-senator to die in poverty on a back street? Yet by the end of the century the young Cornish Americans were venturing into Arizona, undismayed by warnings that it was "the land of rattlesnakes and Plutonian temperatures", by rumours of bitter wars with the Apaches and by reports that John Burns Thompson of Grass Valley had died of thirst while prospecting on Mule Mountain. After all, tragedies were expected almost hourly in every mining community, and Grass Valley had its fair share. Peter Werry died of consumption before he was 30; Peter James was drowned in a forgotten boghole on an old hydraulic claim near the North Star; and John Crase of Redruth never opened his eyes again after a trepanning operation.

Of the lives of these miners little is known for they appear in print only when they marry, have children, pass away or occupy some public office. William George was born in Roche in 1840, exchanged the wildness of its moorland for California when he was nineteen, was elected three times Mayor of Grass Valley, and before he was forty was returned as a Republican candidate to the Californian Senate[1]. R. H. Williams, who reached Grass Valley from Australia, worked steadily in the mines for twenty-two years, retired at the age of fifty-two to begin a new life as a merchant and became one of the town's trustees. P. H. Paynter from Mineral Point by way of Panama and Sutter's Creek held the record for length of service as a magistrate for Nevada County

---

[1] Special edition of *The Grass Valley Union* (1895).

and was the first president of the Miners' Union; and his successor was M. W. Argall, a blacksmith from the North Star and the New York Hill mines. One Cousin Jack gave his life for the community, William H. Pascoe. Sheriff and Town Marshal he was shot through the heart at 10 o'clock on the night of 30 June 1893 on a sidewalk in Grass Valley by the gunman he was looking for[1]. And some survived to become world-wide authorities on mining. Charles Stocks, after posts in New York, New Jersey and Pennsylvania before arriving in Grass Valley in 1872 where he was foreman at the Empire and the North Star, became a consultant for London companies, one of which despatched him on a tour of Europe to study mining methods. But the most widely known and most affectionately regarded was James Bennallack who, as a young man of twenty, had worked his way northwards along the Mother Lode from Sonora. Foreman of the New York Hill, the North Star and Empire, and then superintendent of the Gold Point and the Central North Star, he too was a consultant for London capitalists, his work taking him to Colorado, Idaho, Arizona and New Mexico.

Others displayed a natural aptitude for business. John Glasson, a mere moulder in Nevada City, shrewdly invested in the Imperial, the Phoenix, the North Banner and the North Star mines, and was so successful that he bought out the Grass Valley Gas Company. John Thomas, after 18 years underground by day and many nights of study at school, finally arrived at the position of manager of the Citizens' Bank in Nevada City; moreover, he was elected ten times deputy-recorder for the county. W. J. Mitchell forsook mining for butchering and at the age of forty-six filled out his time as bill-clerk to the Californian Legislature at Sacramento. William and Charles Prisk bought a dying *Daily Union* newspaper, modernized it with a new telegraphic service, sold it at a profit and in 1900 captured the reading public of Long Beach and Pasadena with their two papers, *The Star-News* and *The Press-Telegram*, both of which were renowned for their high standards of reporting[2].

But most of the Grass Valley Cousin Jacks are remembered now only as the chisel cuts of a monumental mason on a slab of granite in the sunlight over Boot Hill. William Cock leaves the grit and dirt of Redruth in 1873, works in New Jersey alone until he has saved enough money to pay for his wife's passage, raises a family in the Nevada wilderness of Pioche and then transports them to the gentler slopes of

---

1   Edmund Kinyon, *Factual Narratives of the Counties of Nevada, Placer, Sierra, Yuba, and Portions of Plumas and Butte* (Grass Valley : 1949), pp. 72-3.
2   Ibid., pp. 164-5.

Grass Valley. John Caddy from Helston in 1866 invests in the Mary Emma Quartz Mining Company at Eureka, crosses the Sierras during the depressions of the early 1870's, becomes an American citizen, returns to Cornwall to manage his father's farm for ten years, marries the daughter of a Nancledra blacksmith and takes her to Grass Valley, and then dies of cholera a few weeks after their arrival. But of others there are only the most meagre facts: Clarence Jenkins leaving Marazion in 1884 and William Henry Bone displaying his skill as a driller at Smartsville in 1868.

Stephen Thomas was fortunate to have a daughter who read history at the University of California and was a librarian at Oakland, for she has kept a careful record of all his movements[1]. His family had mined for generations at Porthtowan where today seagulls scream round the gaunt chimneys of engine houses long since abandoned and meadow larks sing high above stony valleys that drop down to the sea, and where the mines they knew so well, Wheal Charlotte, Wheal Towan and Wheal Lushington, are no more than the crumbled ruins of a copper-girt fortress that once dominated a thriving community. His father was Richard Thomas who, in the depressed 1830's, elected to take a chance in Cuba rather than Wisconsin and then returned home to marry a widow. Their three sons, with mining in their bones and poverty as their lot, all emigrated to America, led by the eldest, William, who foraged in the coalfields of eastern Pennsylvania with his cousin Cyrus Penrose. Stephen, with a ticket paid for by his brother, prospected for many years at high altitudes in Colorado, living in rough cabins above the timber line and once narrowly missing death by starvation. With his father's copy of Bunyan's *The Pilgrim's Progress* in his pocket he arrived in Virginia City in 1873 just at the time when James Fair, the superintendent of the Consolidated Virginia, was dismissing Cousin Jacks wholesale because "these damned Cornishmen know too much". Disappointed with Nevada he crossed the Atlantic in June 1875 to visit his parents, returned to Pennsylvania, worked in a store, married an American and settled her in Grass Valley near his two brothers. Here he stayed for seven years, brightening the lives of many a Cousin Jack with his cornet and flute and the melodies he piped from the *Companion to the Wesleyan Hymn Book* and his own collection of tunes he transcribed and preserved in a book across the flyleaf of which he wrote: "Stephen Thomas His Music Book, Porth-Towan January 5 1865". His enthusiasm for making music was so infectious that he soon

---

[1] Miss Mabel Thomas, to whom I am indebted over many years for the narrative about her father. She died in April 1965.

gathered other instrumentalists, eager to practise their harmonies across his dining-room table; and in 1875 he formed the Thomas Silver Cornet Band with the object of reviving the singing of carols at Christmas.

His reputation along the Mother Lode as a miner, consultant and handler of men was high; "never turn away a man carrying blankets", was one of his well remembered sayings. So no one was surprised when in 1889 he was invited to be foreman of the Sierra Buttes mine, twelve miles above Downieville and at the forbidding altitude of over 9,000 ft. The condition of the mine was uninviting; it was losing money; no new veins had been discovered; a summer fire had destroyed the upper workings; and snow slides had carried away one of the mills. It was in such poor shape that he could have bought it outright if only he had the capital. However he guaranteed he would make it pay if the management would make him superintendent. Assisted by only twenty miners, instead of the normal 300, he did all the surveying and assaying, managed the boarding houses, paid the bills and wages, kept the accounts and wrote reports, directed the milling and retorting of the bullion, and even rode shot-gun down the trails. Conditions were often trying as when the entire seven miles of flumes bringing water to the mine from the Sardine Lakes would be choked with snow at an altitude of 5,500 ft. So great a responsibility, however, taxed even his strength and after five years forced him to leave the mountains for sea-level at Oakland, where he was employed as a consultant until the gold fever of Alaska threw him unto a cold sweat of excitement. Resist this he could not and in June 1898 he joined a "Gold Hunter Party" outward bound from San Francisco.

It proved to be a disastrous undertaking and he was lucky to see his family again. They were late in the year steaming up the Yukon and they met a barrier of ice that forced them to turn back and then abandon ship. A pencilled entry in his diary reveals the misery of these days:

March 31, 1899. This is my fiftieth Birthday. Oh, what years of toil they have been and now with a body almost worn out I am in ice-closed Alaska, looking for a fortune, but it seems further away than ever. All hope of success in this venture is abandoned by me.

For twenty-two days he endured the agonies of drifting in an open boat with no shelter from strong winds and excessive rain until he was stricken with pneumonia. It was a miracle that he survived for he was already riddled with silicosis and he died four years later in Oakland.

Other Cornishmen had worked their way up the canyons of the Feather River, Rio de las Plumas to Louis Argello who discovered it in 1820. But by the end of the century they had all been scattered across the mountains and deserts to the rolling Pacific like the feathers blown upon the river or the dying fall of the last notes of *Airs for the Cornet* of Stephen Thomas.

.   .   .

THE NEW ALMADEN QUICKSILVER MINE

To an Englishman from Cornwall, travelling south from San Francisco is as exhilarating an experience as walking a Roman road through the heart of his own country, for here is one of the historic highways of the New World: El Camino Real, the Royal Road, along which the Jesuit Fathers and the Spanish soldiers pressed northwards, founding their missions and their pueblos at the very time when the thirteen colonies on the east coast were severing their connection with England for all time. This is America's land of saints and of settlements dedicated to San Bruno, San Mateo, San Carlos, Santa Clara, San José and a host of others; and it was the same Iberian people, though of an earlier time, that also made Cornwall a land of saints, with its villages named after St. Buryan, St. Cleer, St. Michael and St. Anthony. Santa Clara's valley of walnuts, apricots, peaches and prunes, flanked by the golden cones of the Livermore Hills and the wooded slopes of the Coast Range, has another connection with the south-west of England, for it was here that Sir Francis Drake made an English gesture in the name of Queen Elizabeth the First and claimed it as New England; and so it remained on some maps until as late as 1834.

In the eighteenth century, when the sick body of imperial Spain lay dying in Europe, its soul went marching on to the shores of the Pacific to preside over the religion and the culture of a new world. Almost a hundred years later yet another link was forged between the old and the new, but this time one of rare and precious metal. New Spain belongs to the era of seamen and conquistadores; New Almaden to the age of furnaces white hot in the service of man. It lies ten miles south of the mission and pueblo of St. Joseph and has the double distinction of being the first mine to be registered in California and the first mine of its kind on the North American continent, for in its depths fabled red cinnabar deposits imprisoned rare quicksilver. In Spanish, "almaden" simply means "the mine", and there was only one Almaden, that in Spain, until this one was discovered, its rival and finally its superior. Today, no one would imagine that its tremendous hill, now covered with live oak, nature's pall that hides the scars made by rapacious men as they ripped away its trees to build another forest underground, in the hundred years of its working life produced more than 70,000,000 dollars' worth of pure mercury. New Almaden is one of the most historic places in the whole of California, at last recognised as a National Monument; and rightly so, for so rich were its treasures that it broke the monopoly of the English banking house of Rothschild which controlled its namesake in Spain.

That the mine was able to dominate the economics and the politics of a generation which had to decide on the boundaries of slavery and secession was due to the skills of its miners, first Mexican and then Cornish. Among the first to recognize this are three Californians who worship at the shrine of the god Mercury: James Schneider, who is one of the present owners of the mine, manages it after years of neglect, leases claims to miners and works a modern reduction plant; Clyde Arbuckle, the city historian of San José and the curator of the County Museum; and Laurence Bulmore, who was born in a cottage in the Hacienda and whose father was cashier and general agent. One ruling passion governs Bulmore's life since he retired as chief engineer on the last of the sturdy ferry boats that carried tired commuters from San Francisco across the Bay to Oakland before the giant bridges were built: to collect and assemble all the physical remains of New Almaden. And this he is successfully doing with a rare air of dedication, whether at his home in San José, or at the rooms of the New Almaden Historical Society, of which he is the President, located in the Museum on the County Fair Ground. Here may be seen expertly arranged and described Cornish rails, tie-plates and miners' candlesticks; a Mexican mine ladder made by cutting notches in the trunk of a tree; a stop from the organ in the Cornish Methodist Church; the working beam from the Cornish pump that was set up at the Cora Blanca shaft in 1873; as well as specimens of cinnabar and the flasks and scales for weighing the mercury. Most valuable of all his treasures is the remarkable collection of photographs of the mine and its workers made by his father and the mine physician, Dr. S. E. Winn. Some, taken underground by magnesium flares electrically fired, claim to be the very first in the history of mining. And many were used as vital evidence in the famous enquiry of 1887 when Frank J. Sullivan, a candidate for Congress in the Fifth Congressional District of California, brought an action against his victorious rival, Charles N. Felton, that the Mexican and Cornish miners had been intimidated and bribed. The charges were never substantiated and the photographs proved that they were "better fed, better paid, better clothed and better housed" than any other labouring class in California.

The approach to New Almaden is along a road that leads to the Hacienda, the road over which black-plumed horses had once drawn the hearse of many a Cousin Jack for burial at Oakhill Park in San José in the same earth as some of California's heroes like W. H. Eddy. All around is rich ranching country of cattle and fruit that was once owned by these same Cousin Jacks, bought with the cash that they grubbed below ground; and in springtime it is covered with the golden flowers of the acacia or mimosa, as familiar to them then as it is now in Cornwall. At an intersection, near the rusting tracks of the old railroad that once carried the mercury flasks to San José for shipment to San Francisco, is a white board bearing the words "Harry Road" that perpetuates the memory of Captain James Harry, at one time the superintendent of the mine.

The Hacienda still has its level shaded street, lined with trees and cooled by a stream flowing from Alamitos Creek, the creek of the little aspens; and its most conspicuous building is the Casa Grande, the Great House, built in 1854 by Francis Myers under the direction of H. W. Halleck the Civil War veteran, and now the Post Office. Originally intended as a single floor hotel it possessed so gracious a line and air of charm that a second floor was added, and it became not only the residence of the mine manager but a mansion in which to entertain the politicians and bankers of California. It was here that Laurence Bulmore spend his boyhood in those spacious days before 1914, when the house and its lawns took on the languor and serenity of an English vicarage, his sisters and their friends playing croquet or boating on the pool where waterlilies touched the rocks of a garden landscaped by John MacLaren, the architect of Golden Gate Park. Near the Casa Grande stands the adobe brick store, erected in 1848; today it is a saloon, a far cry from the days when, like a village shop in England, it sold almost everything from a lady's hat-pin to a sack of potatoes. Everywhere are historical markers that have been set in position by the New Almaden Historical Society. One is an impressive block of granite on which Cousin Jacks at one time hammered away in their rock drilling contests, dedicated in September 1954 by Clyde Arbuckle on the occasion of the New Almaden Day celebrations.

But at the end of the Hacienda are the first signs of a ghost town that Nature has not yet found time to cover. Where once had been the reduction works with their six giant furnaces to roast the cinnabar now gapes a vast pit. In recent years this has been drained of the last 20,000 dollars' worth of mercury that had escaped into the ground in the old days. Now one solitary building, a shambles of board and brick leaning drunkenly against the wind and rain, is all that remains of the office of Laurence Bulmore's father[1]. A couple of hundred feet above this forlorn guardian of a vanishing past stand two red-brick chimneys, which had once carried away the smoke and fumes along brick ducts and flues built into the hillside. Up on the Hill at Englishtown the boarding house for bachelor Cornishmen has long since vanished, though one on the Hacienda has survived as the Café del Rio under the shadow of the Roman Catholic Church, which Laurence Bulmore's great-aunt, Guadalupe Madera, built as a thanksgiving for the safe return of her son from the Spanish-American War of 1898. But on the Hill are signs of a vigorous regeneration, of woodland slopes coming alive in response to the demands of a modern furnace. Once again the red cinnabar is being brought from the drifts of the Cora Blanca; and once again the lively mercury is flowing from retorts. But only twenty or so miners work leases, whereas at one time there were more than 400. Gone now are their neat trim cottages, once gay with song and dance in celebration of the fiestas of Holy Week, the Fourth of July, and the Sixteenth of September, Mexican Independence Day. Gone too are the Cornish Methodist Church, and the Mexican Church on Graveyard Hill. A round hillock is all that was once the burial ground for the Mexicans[2], and a grassy square the village green of the Cornish.

Giant earthworks silently announce the site of the mighty Randol and Harry shafts; and what appears to be a granite fortress of Spanish origin with emplacements for guns to repel the ubiquitous Elizabethan seaman is in reality the old engine house of the Buena Vista shaft. Here was the mechanical show-piece of New Almaden, a Cornish beam-engine with a flywheel 25 ft in diameter and weighing 25 tons that drove a pump which could raise 315 gallons of water a minute from a depth of 3,000 ft.[3] Now not a scrap of metal remains except the rusting iron rods that reinforced the dressed granite. Castellated as it is by the removal of its machinery, Buena Vista, however, will be there overlooking the floor of the Santa Clara Valley long after the old Cornish schoolhouse has disappeared which, defiant of wind, rain and the sun, is poised on the very edge of total collapse. Now its inside is a desert of twisted and broken laths and beams, its only occupants an occasional rattlesnake or a swarm of wasps[4].

---

[1] This was demolished in 1965.
[2] Spanish Camp has now entirely disappeared.
[3] J. Schneider, 'The famous New Almaden Mine', *The Pony Express* (January-March, 1950).
[4] The Cornish schoolhouse has disappeared since this chapter was written.

New Almaden, producer of more than a third of all the mercury in the United States, has had a chequered and convulsive life. Geologists have described its mineral belt as composed of complex Franciscan (Jurassic) rocks, folded and intruded by sills of serpentine[1]; but its outcropping of heavy red ore, known as cinnabar, was known to the Indians long before white men set foot in California. The "vermilion cave" in Los Capitancillos, the "little captains", was where the Olhones or Costanes Indians had adorned their bodies with the red pigment for many years but it was not until 1824 that the source of this ritual was discovered. Two Spaniards, Antonio Suñol and Luis Chabolla, visited the cave and mistakenly thought the cinnabar contained silver, so their efforts to extract it with the aid of a crude arrastra on the banks of Los Alamitos Creek ended in failure. This is not surprising for the lure of silver blinded most Spaniards to the existence of other metals, so it was left to the Mexicans to discover the mercury after Spain had relinquished its feeble hold on California.

In 1845 Andres Castillero, a captain in the Mexican army who was visiting the mission at Santa Clara, heard of a red ore that was being used for the decoration of the chapel. He tested it for quicksilver by the crude but, as it happened, the only method: distillation. The story is that he pulverised the ore, threw it on live coals, then picked up a tumbler of water, threw the water on the coals, inverted the tumbler to catch the vapour, and saw attached to its sides tiny globules of quicksilver. Learning of the source of the red ore, he filed a claim with the Alcalde of the Pueblo San José de Guadaloupe and measured out his mine to a distance of 9,000 ft in every direction from the mouth of the cave. He called his mine the Santa Clara and brought William Chard from New York to operate it. But they were not at all successful in spite of Chard's ingenious use of gun-barrels and cast-iron whaling kettles for the distillation process. So when California became American in 1846, Castillero and his twenty-three partners decided to sell out to the Anglo-Mexican corporation of Barron, Forbes and Company of Tepic, Mexico.

The transaction was to have a profound effect on the future of the mine for the new company bought it in order to free their gold from its reliance on the expensive Spanish quicksilver for refining. Its name was changed and, as a gesture of defiance to the Rothschilds, Santa Clara became *New* Almaden. Equipment and miners were brought

---

[1] See E. H. Bailey, *Suggestions for the Exploration of the New Almaden Quicksilver Mine* (United States Division of Mines: 1951).

from Mexico under one of the most experienced of captains, Henry Halleck, who supervised the erection of five furnaces and condensers; and in 1851 no less than 5,000,000 lbs of cinnabar were roasted to yield 20,000 flasks of mercury. But this very success became a temptation and embarrassment, for it triggered off a whole series of law-suits about the ownership of the mine, some shafts being on the land of one claimant, the Hacienda on that of another. Litigation dragged on for twelve years, uncertainty set a brake on production, and for two years no mining took place at all. For many reasons, which have never been satisfactorily explained, a final decision went against the company in the Supreme Court by a majority of three to two at the very time that a new company was being organised in New York to take over its assets and terminate English financial interests in this part of California; so in 1863 the New Almaden Company was forced to sell for the relatively meagre sum of 1,750,000 dollars.

The new company, the all-American Quicksilver Mining Company, soon ran into difficulties, for it seemed unfortunate or inept in its choice of superintendent, since none of them evinced any understanding of the Mexicans. James A. Nowland (1865-7) bullied them with two six-shooters strapped to his waist, while his successor, E. J. Mayo (1867-70), was more interested in ensuring that they voted Democrat than in extracting mercury. Moreover, real mining experience was needed if the mine was to maintain its lead over its Spanish rival for there were serious limitations to the capacities of the Mexicans since, apart from their native lack of drive, their methods were wasteful and inefficient. Their usual drill was to cross-out the cinnabar veins and then follow the indications of smaller ones, called "tribos", that often led to larger deposits, until the mine looked like a rabbit warren. Their system of extracting the ore and raising it to the surface was also most primitive, notched tree trunks serving as ladders up which the miner clambered, carrying on his back about 200 lbs of ore in an ox-hide bag fastened by a strap to his forehead.

It is generally supposed that it was in 1870, when James B. Randol was appointed manager, that the company seriously began to consider replacing the Mexicans with Cornish miners, but there is ample evidence that they were at New Almaden some time earlier, their reputation having preceded them from Grass Valley and other mines on the Mother Lode. In 1857 Sherman Day, the civil engineer at the mine, complaining to his father, the President of Yale, about the Mexican pre-occupation with bull-fights and fiestas at the Feast of the Holy Cross, remarked: "We are gradually getting rid of these Mexican miners and substituting Cornish miners in their stead"[1]. In the following year he reported: "The mining operations here are on a larger scale than any other I have known in California. There is already a tunnel of about 1,000 feet long, with rail track on it—an underground railroad". There can be no doubt that this was the work of Cousin Jacks for the laying of a tramroad was one of the innovations they brought from Cornwall. Moreover, they soon displayed how expert they were in nosing out the next hiding place of the elusive quicksilver. Instead of the hit and miss methods of the Mexican the Cornishmen would dig test pits and trenches along a lode and, if the vein seemed profitable, continue the exploration by sinking a shaft. This was done by boring holes in the rock, inserting charges of gunpowder, blasting it and then hoisting the loose rock to the surface in buckets or skips, timbering or tulling as they went. When they reached a certain depth, they cut out a level and began drifting along it, mining the ore in chambers which were called stopes, and allowing the water to run into a sump which was then pumped to the surface[2]. Compared with the best Mexican achievements these methods, standard in Cornwall for generations, were quite revolutionary.

How many Cornish worked at New Almaden before 1870 is a matter of conjecture. Between shifts or at weekends they seem to have patronised a saloon on First Street in San José, kept by an Englishman, Charley Barr. There is a tale that they were incensed by an article in *The San José Independent* written by Major H. S. Foote who alleged that citizens were afraid to pass the cemetery at night because the ghost of one of their dead cousins was abroad. When they accidently met Foote at Barr's they threatened to beat him up and were only restrained from doing so, when Foote agreed to drink to the ghost's health[3]. One of them may well have been William L. Pearce who could not remember where he was born but vaguely thought the place to be "twenty miles east of Land's End", that is, somewhere in the neighbourhood of Redruth and Camborne, though as a boy, one of eleven children, he worked underground at old Balleswidden mine near St. Just. He emigrated to America in 1848, working first in the Bristol copper mines in Connecticut, then in the lead mines of Wisconsin and later in the gold mines of William Creek, British Columbia. It is not clear when he appeared in New Almaden, but he declares that he spent

1 Bulmore Mss., letter of 2 May 1857.
2 Harries, p. 82.
3 Ed. Eugene T. Sawyer, *The History of Santa Clara County* (Los Angeles: 1922), p. 115.

fifteen years there and in 1869 was rich enough to buy a ranch at Los Gatos[1]. But beyond these facts it is not easy to identify him further. Was he the Captain Pearce, one of Randol's underground men, who was given a bonus of 10 cents per flask of mercury that led to so much wrangling among the other miners that he had to be dismissed[2]? Or was he the husband of the Cornishwoman who appears in one of the company's letter books thus: "Please procure to Mrs. Mary Pearce, East End, Redruth, Cornwall, England, a draft for £10 stg."? The identity of both of them, however, is less important than the fact that money was being remitted from New Almaden to Cornwall in 1867, for a further letter shows that she was receiving sums at regular intervals, enough to feed her and her family when there was widespread poverty and unemployment in Cornwall following the collapse of many country banks in 1866[3]. It is not unreasonable to suppose that she would tell her neighbours of the steady work that awaited their mining relatives if they would only make the long voyage to New Almaden. Perhaps this was preferable to Grass Valley because of its easy access from San Fancisco, for as one of the engineers observed in 1865: "From the mine-ridge to the north west the view is unbroken to the mountains of Marin County beyond San Francisco. The City is hid by projecting hills, but Angel Island and the waters of the Bay are distinctly visible . . . Occasionally the smoky train of the ferry boats can be seen floating away over the waters and the fleecy clouds streaming through the valley show where the iron horse is rushing over the land"[4].

Another Cousin Jack who enjoyed this view in the pre-Randol era was Arthur Berryman[5]. Of a roving and restless disposition he left Cornwall in 1854 when he was twenty and wandered around camps at Sykesville in Maryland, Minersville in Pennsylvania, and then Chile, Bolivia and Peru before settling at New Almaden, where he stayed for seven years until the gold fever gripped him again and he was off to British Columbia. There he seems to have been lucky, for he was back within a year and in 1882 retired from mining to become the owner of the Los Gatos Hotel on the stage route between Santa Cruz and San José; and so successful was he that within five years he sold it for a handsome profit and entered the real estate market. One of his contemporaries was Richard Bailey Harper who, born at Redruth in 1847,

1 Ed. H. S. Foote, *Pen Pictures from the Garden of the World* (Chicago: 1888).
2 Bulmore Mss.
3 Schneider Mss., letter from J. E. Mayo to S. F. Butterworth, 23 September 1867.
4 Bulmore Mss., report of engineer C. E. Hawley, May 1865, in a letter book of the mine.
5 Foote.

voyaged to San Francisco as a boy of seventeen by the long sea route round Cape Horn; after a year on the Mother Lode he was in New Almaden in 1865, but left for Mexico before the coming of Randol, ultimately returning to manage the Santa Teresa quicksilver mine at San José[1]. Then there was the whole family of the Geaches from Tywardreath, though we know only a little of one of them, John, who was a boy of eleven when his parents in 1862 took the stage round the curve of the mountain to their cottage on the Hill, for he married a daughter of one of the greatest mine captains of them all, James Harry, of whom mention has already been made. Second only to Harry was James Varcoe, also from the same district as the Geaches, a giant of a man who, although his best work was done for Randol, was first employed by Mayo in 1867 on the strength of ten years of accumulated experience in New Jersey, Michigan, Grass Valley, Virginia City and Fresno; he too took unto himself a Cornish wife, Louisa, from the Rowes already settled on the Hill. An excellent mine captain, he also played a significant part in the social life of the community, being a member of the San José Lodge No. 34 of the Independent Order of Oddfellows and of the Mount Hamilton Lodge of the Ancient Order of United Workmen, as well as a trustee of the Benevolent Society of New Almaden[2]. His home was still there when he died at Pacific Grove in June 1895; and his grand-daughter Elizabeth Colliver, whose father Charles Tonkin married Varcoe's daughter Anna, still lives in San José.

There were perhaps 30 or more Cousin Jacks with their families between 1860 and 1870 for the census of 1860 enumerates 44 miners described as "English". The 1870 census, however, shows that they had increased to 70 English miners, 31 English women and 20 English children, numbers which were to double in the next ten years or so as part of a deliberate policy. Randol's appointment as manager in 1870 was the company's answer to the critical state of their finances for bankruptcy seemed only just round the corner. The reason was not the shortage of cinnabar, but the problem of where to find new deposits; and this seemed to be beyond the capacity of Mayo and his Mexicans, captained by the few Cornish he had happened on. The ore bodies were large and exceptionally rich, the largest being almost 200 ft wide, 15 ft thick and extending to a depth of almost 1500 ft, producing the exceptional output of a flask of mercury to a ton of ore[3]. But by 1870

1 Bulmore Mss.
2 Foote.
3 Bailey, p. 263.

the existing crews had only reached the 800 ft level, the Day tunnel had not been connected to the main shaft and there was only one engine working underground. The result was disastrous; output had fallen from 5,000 flasks a month in 1866 to 750 in 1869 and the value of shares from 100 dollars to 11 cents; there was an accumulated debt of almost 2,000,000 dollars; the Bank of California was owed another 250,000 dollars; and a further 200,000 dollars had to be raised by private loans to keep the mine going[1]. At first sight Randol seemed totally unsuited to grappling with this problem; he was not a mining engineer; and it is said that he never went underground once during the twenty years he was in charge of New Almaden. But he was not yet thirty, and his most valuable quality was his understanding of men. He soon realised that the solution to his problem lay in increasing the supply of Cousin Jacks for he appreciated their skill, ability and experience; and time proved him to be right for, when he retired in 1892, all debts had been honoured and 3,000,000 dollars paid out to shareholders.

Most of Randol's Cornish recruits were hired along the Mother Lode and enticed from the camps in Nevada County without much difficulty, for he know many of them personally at the North Bloom-field mine in which he possessed financial interests; but he also advertised in San Francisco newspapers like *The Alta California* as well as *The West Briton*, which was as widely read in Cornwall then as it is now. The result was that by 1884, according to the *San José Directory* of that year, out of a total working force of about 300 the Cornish numbered almost 100 compared with the Mexican 60; moreover, all the Cornish are described as miners, enginemen or captains, whereas only 14 of the Mexican are miners, the remainder being teamsters and labourers. Three years later the Cornish super-intendent Hennen Jennings, in his evidence at the enquiry about the rigging of elections, reckoned that the total number of workers was between 450 and 460, of which the Mexicans accounted for 120 and the "English from different parts of England" 181. The remainder consisted of Americans, among whom were classed descendants of Mexicans and English born in the United States, Spaniards, Swiss, Swedes, Norwegians, Russians, Italians, Dutch, Irish, Austrians and "some of unknown nationality". The three dominant groups were the English, the Mexicans and the Americans; but it was the English who accounted for rather more than half of the working population,

---

[1] 'A Contested Election in California', reprinted from *The San José Mercury* (1887), pp. 7-8.

and among these must be included a high proportion of Cornish who occupied the most responsible positions. Not many more miners seem to have been engaged after 1887 though, again according to the *San José Directory* of 1889, the entire population had increased to about 1,400, proof of the stable conditions that prevailed at New Almaden and the rightness of Randol's policy of preferring married miners, for here was a balanced community centred round a single mine operated by a single company.

Now one of the criticisms that Americans usually levelled at immigrants was that they tended to depress wages, but this was never the case at New Almaden because Cornish skills, in continuous demand and therefore commanding their own price, raised the wages and the standard of living of all with whom they worked. For instance, in 1889 a blacksmith could earn as much as 60 dollars a month, a shaftsman 70 dollars and a pumpman 90 dollars; timbermen on a daily rate were receiving from 2 to 2.50 dollars, trammers from 1.75 to 2.25 dollars, skip fillers 2 dollars and underground labourers between 1.50 and 2 dollars[1].

These high rates of pay for all workers grew out of the Cornish contract system of wages which placed a premium on the best worker. Instead of being paid a flat daily or weekly wage Cousin Jacks would bid against each other to drive so many yards into a tunnel, drift, winze or shaft for a certain price, payment then being made according to the number of yards achieved. Notices of contracts were displayed in advance outside the company offices and even advertised in the San Francisco and San José newspapers; while the miners themselves were always allowed to inspect the work before they made their bids. These had to be made in writing, signed by the mine captain or surveyor and then tabulated in a book, so that the lowest or most advantageous bid could easily be picked out. The miners also had to sign an undertaking on a special form that read: "Six full shifts must be worked weekly or a fine of one dollar be paid for each shift lost, the contractors to do all the necessary timbering"[2].

The Cornish contract system worked so well that the company extended it to include almost every kind of manual labour except such special skills as those of pumpman and engineer. There were in fact six classes of contract at New Almaden. The first and most important, already described, was contract by yardage. The second was that of

---

[1] J. B. Randol, *Quicksilver, an Extract from a Report on the Mineral Industries in U.S. at the Eleventh Census, 1890* (Washington: 1892), p. 241.
[2] Bulmore Mss.

"drilling by the foot", where the men were paid by the length of the drill hole they drove into an ore chamber under the direction of the shift boss. The third was "tributing"; here the contractor was paid for the amount of ore he blasted and excavated, doing the timbering as he went along. The fourth related to tramming, the trammers being paid for the number of tons of ore they loaded in cars and shifted from a level to an outlet tunnel. The skip fillers were likewise paid by contract for the number of tons of ore they shovelled into the skips and hoisted to the surface. Finally the teamsters contracted to haul the ore from the planillas, or the dressing floors, over the roads to the chutes, where it travelled up an inclined railway to the furnaces on the Hacienda; and even the mechanics on the incline were included in the system. Its recognised merit was described thus: "Under the contract system there is natural selection and weeding out going on among employees, the less competent miners and dissolute miners not being able to compete with the more skilled and better class of miners"[1].

The Mexicans no doubt were perplexed by the complexity of the wage structure and the fact that the best contracts always went to the Cornish but they shared in the general efficiency all the same and learned new techniques of drilling and blasting. The Cousin Jack would first start a hole in the rock with his pick, then his mate would hold the drill in the hole while he wielded the hammer, turning it deftly and swiftly with each blow, using a larger drill until the hole was four inches deep. They showed the Mexicans the need for keeping the holes clean, using a spoon to remove the rock cuttings and the sludge, and drying them out with a piece of cloth before inserting the charge of gunpowder, for they used water to facilitate drilling in "down holes"[2]. No less fascinating to watch was their dexterity with square-set timbering as they worked down a shaft, or their efficiency at securing the roof with posts, caps and laggings as they excavated a stope; they would also use waste rock to build up walls as a protection against pressures[3]. In fact they were so good that Randol reckoned that they could perform the impossible task of drifting through muddy waters. For two other innovations the Mexicans must have been specially grateful, reducing as they did their heavy load of miserable toil: the Cornish wheelbarrow with its wooden sides which the tenateros called "Jack Wheelbarrow"; and the tramway with its flat iron rails, measuring ½in by 3 in, and secured by iron wedges driven in a slot.

[1] 'A Contested Election etc.', pp. 7-8.
[2] Harries, p. 88.
[3] Harries, p. 95.

The Cousin Jacks left their impression on the Hill as formidably as they had done in Grass Valley: the mighty Randol shaft, cribbed with massive 12 in timbers, and then the Washington, the Buena Vista, the American, the St. George, the Harry, the Victoria and, deepest of them all, the Santa Ysabel. Among them none was more outstanding than Captain James Harry who, though he had little formal education in Cornwall, could walk round a heap of cinnabar and accurately estimate its weight or throw a straw into Alamitos Creek and expertly calculate its hourly flow. Born in Breage in 1833, a village older by a thousand years than any Spanish settlement in California, he slaved as a boy at Wheal Vor, "the mine of the furrow", and witnessed his father brought up from the darkness to be buried. He had five brothers and they all, one by one, made America their home; James was thirty-three and married to Elizabeth Carlyon when he decided to join his brother Joseph in Grass Valley but, after two years, became one of Randol's men at New Almaden. At first he was employed on skilled contract work in the drifts, sometimes earning as little as 3 dollars a week and sometimes as much as 35 dollars; or timbering at 3 dollars a day and installing pumps at 4. His great moment came when Randol entrusted him with finishing the shaft that bore his name; it had been started by the two Cornish captains, Grey and Pearce, in June 1871 some six months after his arrival, but had run into unexpected difficulties. When Harry had finished with it, the Randol was a masterpiece of engineering and possessed so brilliant a reputation that for many years it blinded the management to the possibilities of cinnabar deposits elsewhere.

He also improved the mechanics for raising the ore; instead of the bucket, which could carry only 500 lbs and which swung dangerously round the rope with which it was hauled to the surface by a steam winch, he installed a Cornish skip. This was an iron box with a door in front, capable of discharging a load of nearly two tons at a time. From now onwards Harry's promotion was rapid. Shift boss in 1873, foreman of the Cora Blanca in 1876, in charge of sinking the Ysabel in 1877 (a record year of 80,000 flasks), four years later he was senior mine captain and second only to the superintendent: "I carry out the work underground, running drifts, sinking shafts, and raising and sinking winzes, and seeing that the mine is kept properly timbered, well ventilated, and to do everything that we know for the safety of the men in regard to the timbering, and to take out the ore in the cheapest way, putting in shoots to conduct the ore from level to level and to transport it to the shaft in cars, and to work the mine as econo-

FORSYTH LIBRARY
FORT HAYS KANSAS STATE COLLEGE

mically as we can"[1]. Then in 1889, when it was thought that New Almaden was exhausted, he made fresh surveys, uncovered the tiny Santa Maria shaft, enlarged it and called it by his own name, and so found a new ore body that kept the mine busy for another five years. It was natural therefore, when Randol retired in 1892, that Harry should follow him as superintendent and manager, receiving a salary which in Cornwall then must have seemed astronomical, 250 dollars a month[2]. Prudently he invested his income in land, buying company property to the extent of 300 acres, equivalent in Cornwall to the size of four farms[3]. And when production fell in the 1890's, he boldly bid to discover new ore bodies by driving the 800 ft level of Deep Gulch tunnel until it ran under the Harry. Here his Cornishmen could begin to raise and then drift east, where he reckoned the next Aladdin's cave would be found[4].

Harry's *Time Book* of 1885 gives the names of some of the 300 Cousin Jacks who worked under him. At the Buena Vista the engineers were John Bishop, John Eddy, Charles and John Pearce, and their assistants Richard Pearce, the fireman and William Jacka, the mechanic; the miners were Alf Tregonning, Richard Pearce Snr. and Thomas Trevarthan. At the Randol the brothers Thomas and Charles Tonkin were operating the hoists with a company of mechanics that included John Edwards, John Robins, Harry Ralph and Joseph Pearce, while the "helpers" were John Bowden, Richard Jacka, Thomas Martin, George Reseigh and William Bunney Jr. Two more Tonkinses were keeping the wheels turning over at the Ysabel, John and William, with the assistance of John Johns and the two firemen, Charles Harry and Percy Dunstan. Timbermen included James Varcoe, John Martin, T. B. Cornish, William Gilbert, Charles Harris, Thomas Matthews, Thomas Pryor, John Rowe, William Tonkin, J. O. Pearce and William Pearce, while the blasters were William Bunney, Thomas Odgers, Samuel Pearce, John Reseigh and William Doidge, once chosen specially for his photograph as a typical Cousin Jack.

Some others survive in the 1879 admissions register of the Methodist Church[5]: James Johns, Thomas Matthews, Thomas Pascoe, James Trengrove, William Curnow, Thomas Henwood, Samuel Pearce and Alfred Tregonning. Pictures of the latter appeared on many a glossy

---

1 'A Contested Election etc.', pp. 128.
2 Bulmore Mss.
3 Ibid., letter from Company acknowledging his request for a further 25 acres, and an undated map, showing that Harry already possessed 292 acres.
4 Ibid., Harry's report on New Almaden, 17 May 1895.
5 Bulmore Mss.

brochure, showing him pouring the mercury into flasks, his sleeves rolled up to the elbows and a leather apron around his waist. Others remain only fleeting memories. William Pearce one morning "arose, got breakfast for his sick wife and said you had better have a cup of tea, and at once closed his eyes for ever". James Rule, a wasted victim of consumption, died "in great peace". Charles Harris in 1891 was killed when a ton of rock fell on him as he was replacing some defective timbers. Jim Prout felt the whole earth around him shudder as the deadly San Andreas fault shifted and wrecked San Francisco. Of the three Bartle brothers from Beacon near Camborne, Harry disappeared without trace, Frank declared his American citizenship in Virginia City in 1874, and John finally left New Almaden during the pre-1914 depression and was killed by a train. T. B. Cornish watched two of his Cousin Jack friends perish of the bubonic plague in Cuba. John George from Perranarworthal had mined at Pilot Knob in Missouri and Battle Mountain in Nevada before arriving at New Almaden. Charles Berryman and his wife chose New Almaden after they grew weary of the bitter winters round the iron harbour of Marquette in North Michigan, while their son shrugged mining altogether for a career as a sanitary engineer at Los Gatos where he died in 1963. All too soon the links with Cornwall snap as the events of their lives fall away into the shadows; as in the case of Martin Oates and Frank Kessel from Chacewater; James Hodge from Nancledra and his wife from Treswithian Downs; and Peter Barrett, now no more than a name. And yet William J. Trevorrow, in his retirement, returned time and time again to the yellow gorse of his childhood at Par, and was admitted into the Cornish Gorsedd as a bard with the title of Lef An Howlsedhas Pel, "the Voice of the Sunset from the other side of the World". Edwin Willoughby is remembered as the superintendent of the Soledad mine in Mexico, local preacher at New Almaden, and for thirty years as assistant to S. B. Christy, the Professor of Mining and Metallurgy in the University of California at Berkeley; Frank Argall as a driller of granite, and of teeth too, after training as a dental surgeon at the University of the Pacific; and Joseph Hancock as a member of the Board of Education at San José, raising the cash himself to study at Stanford University after he had seen his father killed in the mine.

New Almaden was always a romantic place for the visitor. One went so far as to exclaim that it was "an earthly Elysium", where the lemon verbena grew in the open to a height of 20 ft. Another remarked on its haunting attractiveness under a full moon with "the winding trails with the black masses of shadow from the clumps of live-oak crossing them, the dark mountain lines rising grandly on every side, the mysterious depths of the canyon, the lights of the Mexican camp scattered over the hills, the close clustered lights of the Cornish camp on the lower range, the wide dim valley below and the far-off barrier of the mountains"[1]. Even in daylight there was something romantic in the sight of Mexicans and Cornish sharing their tortillas and pasties and Fermin Gomez "smoking pipe" with James Goldsworthy; or Cornish hymns competing with the light twang of the Mexican guitar; or fair-skinned Cornish youths hurrying off to the ball with their dark-skinned Mexican girls on Independence Day. But beyond these social occasions there was little mixing by the two races and mixed marriages were not encouraged. Neither was there any hostility. Indeed New Almaden seemed to be a model mining camp where employees wished they had stayed longer or arrived earlier, for they never lost work through a strike or even the threat of a strike.

The main reason for this happy state of affairs was of course the Cornish wage structure, described by Professor S. B. Christy as the "rate for the job" and "by far the best system for attracting the best class of miners", while Hennen Jennings reckoned: "In return for the large pay they could make on their contracts they (the Cornish miners) performed the greatest amount of labour of any set of men I know". The results were high wages for all and high productivity; in 1886-7 the average daily wage for those on yardage contracts was 2.48 dollars and even reached a peak of 5.96 dollars[2]; Jonathan Eddy was earning 85 dollars a month as an engineer, William Doidge 71 dollars as a blaster and James Varcoe 140 dollars as a shaftsman[3]; and there were extras to be won for working in a hot drift or sinking a wet shaft. On the Hill and the Hacienda "there is no such thing as want", many Cousin Jacks invested their savings in the San José Water Company, and more then half of them owned their own houses, which they erected themselves for a ground rent of only 6 dollars a year. Rented houses were available from the Company at a rate of one dollar a month per room; thus Henry Tregonning, a watchman on a monthly salary of 85 dollars a month, rented a four-room cottage for 5 dollars which also included the use of a barn, a yard and a garden. These company houses were not so unprepossessing as one might think; they were constructed of redwood with shingle roofs and board partitions between rooms;

1 Mary Halleck Foote, *A Californian Mining Camp* (Scribners: 1878).
2 'A Contested Election etc.', pp. 13-14, 21, 24. 105-6, 143-4.
3 Bulmore Mss., Time Book for November 1885.

and they were fenced to enclose a garden which the Company stocked with plants. Water was pumped from an adjacent ravine for a nominal rent of 50 cents a month. If to these amenities be added the privilege of pasturing cattle on Company property free of charge, it will be seen that Cornish families for the first time in their lives were enjoying the pleasures of home and garden, displaying so prosperous an air that it was said visitors thought they were at Grace Church on Broadway, New York.

One factor that contributed to the general air of affluence was the company's system of controlled overdrafts at the store, which had the advantage of circumventing indigence and encouraging the thrifty. It arose because pay day, "dia de raya" to the Mexicans, came only every calendar month which imposed a strain on the budgeting capacities of some families, particularly the Mexicans. So that no family should ever be without food, the company therefore issued boletos or tokens, small pieces of cardboard or metal in the usual denominations of money up to a 5 dollar piece, which could be advanced up to the limit of the next month's salary and exchanged for goods in the store. The Cornish were usually more careful than the Mexicans and preferred to see their bank balances growing; but there was no bank on the Hill and so they had to make their deposits at San José, 12 miles away[1]. When they were short of ready cash, they made use of boletos and then soon discovered that they could trade them to the Mexicans for a discount of 20%. One suspects a black market in boletos working to the advantage of the Cornish and tending to impoverish the Mexicans whenever there was a heavy load of personal debt, as for instance in 1890 when the total value of boletos issued was 7,500 dollars[2].

But by and large these shrewd practices were overlooked by the Mexicans who knew they enjoyed the benefits that the Cornish insisted on. None was more influential than their Miners' Fund, a form of health insurance that bound miner to management in a relationship that lasted as long as the mine itself. For a dollar a month a miner and his family received the services of a resident physician, on call at all times of the day and night, and medicines at cost price; confinement cases, however, were regarded as requiring special treatment, and for them a charge of 5 dollars was made, which was ploughed back into the Fund. Its finances, managed by the men themselves, were sound enough to pay the doctor 400 dollars a month, and a secretary and a qualified

---

[1] 'A Contested Election etc.', pp. 28-31, 53, 101-4, 131-4, 153.
[2] J. Schneider 'The New Almaden Mine', *The Pony Express* (August 1949).

druggist 240 dollars each, and to support two hospitals, one for the Mexicans and one for the Cornish. Its benefits were unusually wide for it possessed the power to meet the lodging bills of an injured miner, pay funeral expenses and even allocate grants for the upkeep of graves. The Company's share of the scheme consisted of providing a house for the doctor with an office and a dispensary, and a horse for making his rounds, the most valuable part of his equipment after his instruments for "the resident physician responds to a call at any hour of the day or night, often having dangerous rides on a dark stormy night in winter to visit a labourer's cottage". Dr. W. S. Thorne maintained that in all his thirty years' experience in the mining camps of Nevada and California the sanitary conditions at New Almaden were the best he had ever seen, and the drugs and medicines the finest that could be bought on the west coast.

Training in the rudiments of health seems to have had some effect on curbing carelessness in the mine itself for in 1887 James Harry was boasting that there had not been a fatal accident for three years[1]. Salivation, the only fatal disease known to quicksilver miners, was also well controlled. The Indians had been the first known sufferers from the deadly mercury fumes, released by the sweat of their bodies from the red ore which they used to decorate themselves; the fumes would stimulate the salivary glands, constant dribbling at the mouth would follow, and a brave would die by spitting himself to death. Salivation was always a danger to the Cornish at New Almaden, because there was little chance of controlling the fumes that belched from the chimneys of the reduction plant, but between 1898 and 1903 only nine cases were reported by the physician, Thomas Matthews in 1901 being described as "badly salivated".

An important social amenity that grew from the Miners' Fund was the Helping Hand Club, a benevolent association that started in 1885 to provide single men with some place of amusement and entertainment other than the saloon. As usual there had to be two buildings, one on the Hill for the Mexicans and one on the Hacienda for the Cornish; the latter was an elaborate affair, with seating accommodation for an audience of 400, a raked stage and drop curtains like the Tivoli in San Francisco, and dressing rooms, all provided by the Company. Here touring troupes performed their plays or local amateurs arranged entertainments for some worthy cause, as in 1895 when 85 dollars were collected for Mrs. Lewarne, left destitute with four children after her

---

[1] 'A Contested Election', pp. 8-18, 37-43, 85-7, 133.

husband had been killed in the mine[1]. But the clubs were also adult education centres, for they each had a reading room containing a library of 450 books—"stories, biography and history"—weekly magazines and daily newspapers. The Cousin Jacks it seems were avid readers for they also bought their own papers, more than a hundred being delivered every day on the Hill: *The San Francisco Chronicle* and *The Examiner, The Alta, The Call, The San José Mercury, The Times* and *The Christian Advocate;* many received regularly from overseas their copies of *The Times* and *The West Briton.*

It is not surprising therefore that the standard of education was high. The Company provided three schools, one for the Cornish, another for the Mexicans, and a third for the children of the employees on the Hacienda. There was a staff of one male head teacher and five women, and all seem to have been well paid; for instance, Shumaker, the principal of the Hacienda school, received a salary of 112 dollars a month, which was far in excess of that of his colleagues in England. Attendance at school, of course, was not compulsory; so in 1887, while the total number of children enrolled is 253, the daily average attendance is only 169, the deficiency, it is supposed, being accounted for by the Mexican maxim, "mañana eo otro dia". As far as the Cornish children were concerned there was assuredly not the slightest chance of playing truant from their Methodist Sunday School where they were "better versed in scriptures than most children"[2]. They were also the subject of an interesting experiment in technical education in the summer holidays of 1890 when the Helping Hand Club organised for the school-leavers courses in "plain" cooking, sewing, carpentry and blacksmithing under the direction of James Harry and Angel Delamastro, the Company's master carpenter. The scheme included facilities for the education of their mothers too, for the Company printed, and the Clubs distributed, a booklet entitled *Cookery for Working Men's Wives,* which also contained useful hints on such domestic matters as sanitation in the home and preparations for a visit from the doctor, the whole based on a report of the United States consul in Glasgow who had visited a demonstration school there.

It was beneath this umbrella of social security provided by the Company that the Cornish were enabled to develop their own special voluntary associations. Some belonged to the Sons of St. George, their Lodge perpetuating the name of their Christian hero, General Gordon of Khartoum; others enrolled among the Knights of Pythias,

---

1   *The San José Mercury,* 3 June 1895.
2   'A Contested Election etc.', pp. 17, 51-2, 57, 134.

their lodge appropriately called the Cinnebar; almost all paid their dues to the Independent Order of Oddfellows and the old craft guild of the Ancient Order of United Workmen. But by far the most significant influence on their lives was the Methodist Church, as potent a force as the Mexican fraternal Nuestra Senora de Guadaloupe[1].

As everywhere else, Cornish Methodists were expected to take their churchgoing seriously and there was no room for the sluggard or the doubter. New arrivals at the mine, like Joseph Holman from Johannesburg or William Tonkin from South America, were welcomed to the fold provided they were ready to accept its disciplines, such as regular attendance at the weekly class, where the Bible would be studied as devoutly as at any seminary. Not all could undertake these commitments. In 1879 only eight qualified for admission while between 1879 and 1896 the yearly average was only twenty-five. If they proved irregular in their attendance, they would be "dropped" and notified by a curt letter. "Our church was the salvation of New Almaden", says Mrs. Mary Hodge Hall of Chula Vista, the daughter of James Hodge from Treswithian Downs. Born at Georgetown in Eldorado County and appearing in New Almaden when she was six years old, she recalls that, while the schooling on the Hill was "excellent", there were very few who passed on to the Normal School at San José. She was more fortunate, for she trained as deaconess at the San Francisco Training School where she studied a whole range of subjects to add to her reading of Dickens, Scott and Fennimore Cooper with which she entertained her parents on the long winter evenings at the mine: Christian Evidences, Ethics, Christian History, the History of Methodism, Social Service, Physical Culture, Medicine and Nursing. Thus equipped, she bravely ventured as a social worker into the slum area around Fisherman's Wharf among the drunks and those with a "lowered estimate of themselves", continuing her work of mercy and human reclamation when the earthquake of 1906 destroyed the city.

For those Cousin Jacks who made it their home, New Almaden terminated their existence on a mining frontier; they had reached their little Eldorado, achieved their modest ambitions and worked out for themselves a viable Utopia. Children came like the blossoms in spring in clusters of eight or nine to a family and no mother lost her life in giving life[2]. Divorces were rare, and only one woman ever had to

[1] L. Bulmore, 'The Human Side of New Almaden', *The Pony Express* (Dec. 1949).
[2] Bulmore Mss. in a letter from W. H. Bunney, 14 August 1945. Cornish live births 1876-97 were 300.

An interesting survival (about 1930) of California's mining camp "architecture"; the Miners' Union Hall at Bodie, a notorious silver camp to which the Cousin Jacks flocked from Nevada in the 1880's.
[*Bancroft Library, University of California*]

The Cornish town of Nevada City, California, about 1890, looking down Main Street.
[*Bancroft Library, University of California*]

The young Richard Jose of Lanner, depicted in a contemporary magazine singing his Cornish carols in the streets of Virginia City, Nevada. [Mrs. Therese Jose Hamlin]

A recent photograph of a surviving shelter that Cornish silver miners built at Calico to protect them from the fierce winds that blew across the Mojave Desert.
[Harold O. Weight]

finish out her days in the Poor House. In the Cuban war of 1898 they organised a branch of the Red Cross, Mrs. Geach, Mrs. Pierce, Stella Lanyon and Nellie Tregonning raising funds to buy books, pyjamas and "comfort bags" for the troops in Manila.

Their Cousin Jacks only rarely disturbed the peace of the Hacienda, usually at election time when they voted Republican to keep out the Roman Catholic influence of their Irish rivals[1]. As an industrial élite they did not seek political leadership for a place in society, while their burrowing in the earth denoted a symbolic withdrawal from its tensions and conflicts. Nearer to the globe's core they were more aware of a Divine Presence and hoped that God walked with them there at night when the rock became alive. Unlike the Mexicans they erected no shrine at the bottom of a shaft, yet their hymns cut the blackness no less than the candle flames, reminding them all of the chances and changes of their mortal life. There they would neither dare to whistle nor to kill a rat; robins were sacred; and the mine ponies were their friends, to be remembered at Christmas when they were brought into the sunlight and presented with gifts of apples and sugar[2]. Nor would they work on Good Friday or Innocents' Day when they were reminded of the red hand of persecution. Drinking here was little more than an excuse to visit the Boom-a-Rang, the saloon and boarding house for single men behind the Helping Hand Club. Drilling contests enabled them to display their professional prowess in public; [and their Mountain Echo Band demonstrated the power of their lungs. Led by Joe Bishop, they blew their Cornish, Mexican and American airs through the long canyons, by the waters of Los Alamitos Creek and up the side of the mountain during the Mexican three day fiesta before Ash Wednesday, on Mexican Independence Day, on "Spy Wednesday" in Holy Week when they would hang Judas again, on the Fourth of July, at the annual summer tea "treats" in the church picnic grounds, and at Christmas when a community tree would be decorated in the church and loaded with presents from family to family. Here no tatty travelling theatrical companies, wrestlers and boxers, gamblers and the riff-raff of the mining camps soiled the land during the golden age of the Quicksilver Mining Company; for here were two races, centuries older than the Americans, working and living in harmony. There had been a time when a killing on pay day was not unusual; and once a skeleton had been found in a drift with a bullet hole in its head. James Harry had been warned in

---

1 'A Contested Election etc.', p. 128.
2 Ed. Roderick Peattie, *The Pacific Coast Ranges* (New York: 1946), pp. 161-3.

Grass Valley that New Almaden was a den of robbers and murderers, and the hide-out of the notorious bandit, Tiburcio Vasquez, hanged in San José in 1875; and he never quite forgot his first day on the Hill when he came across a man lying at the side of the road in a pool of blood; he was some woman's third husband and all three of them had been shot dead. That those days of lawlessness had disappeared for ever is due in no small degree to the Cornish. If they knew nothing of the Spanish Main and the exploits of the Devonian Drake along their Californian shores, they proved that the Cornish breed of a later time was no less memorable, if only for their understanding of workers whose names—Inez Acostia, Patricio Andrade, Lazero Castro—seem part of the *dramatis personae* from a romantic comedy of Drake's equally gifted contemporary, Shakespeare.

In 1918 New Almaden was sold for a paltry 125,000 dollars to satisfy its creditors and the giant flywheel of the Buena Vista slowed to a standstill for the last time; the reign of the Quicksilver Mining Company was over, but to the end its mining captain was a Cousin Jack, Richard Harry. On a warm December morning in 1960 the ninety-one years old John Drew from Helston and his bed-ridden wife, a Bishop from Penzance whose father had been hoistman at the Buena Vista, remembered how much they owed to the Harrys for persuading them to leave Grass Valley with eleven children for New Almaden. Drew had started his mining life below grass at the back-breaking task of shovelling fifty tons of cinnabar for the contract price of 3 cents a ton; after that he excelled as a timberman, making drift sets of redwood with a 5 ft saw that he could drive with either hand. On one occasion, when the heavy cumbersome complication of struts and beams was being hauled into position at the mouth of the shaft, it jammed and Drew climbed down a rope to release it; but the rope snapped and Drew and the set fell to the bottom of the shaft; he escaped by a miracle with only a broken shoulder. With his death in 1963 only a handful of survivors in San José live to retell the Cornish saga. Such is Elizabeth Colliver, a grand-daughter of James Varcoe, who recalls her childhood days reading in the Cornish schoolhouse *The Last Days of Pompeii* and *Enoch Arden*. Her Cornish mother-in-law was born in a covered wagon near Lake Tahoe and her father was one of four Tonkin brothers from St. Just who became hoistmen at the Buena Vista. Thomas is remembered for a brave action for, though severely burned when a boiler exploded, he continued to lower the cage and so saved the lives of several miners. John followed Hennen Jennings to the El Calleo mine in Venezuela where he was outstanding "in the capacity of timberman, shift boss and foreman of shaft sinking, to the entire satisfaction of the officers of the company"[1]. His daughter, married to a Californian State Senator, now takes the long trail back to St. Just, where she is at once recognized because "she looks like a Tonkin".

.   .   .

THE SOUTHERN MINES, THE MOJAVE DESERT, DEATH VALLEY AND YOSEMITE

THE heartland of the southern mines is Tuolumne County, where Indians once dwelt in caves and miners worked the quartz lodes round Soulsbyville and the rich river beds round Jacksonville. To reach it from its neighbour, Calaveras County, the land of skulls, the Stanislaus River must first be crossed (the river in which beaver was trapped by Russians in the days before gold), either at Robinson's Ferry or at Knight's Ferry with its wooden canopied bridge built by General Grant in 1864. The road from here to Sonora gives a spectacular view of the quartz vein of the Mother Lode clearly exposed near the Harvard mine and looking like a sandwich filling of cream on the mountain side. The geologist will explain this phenomenon in terms of an extremely complicated sequence of uplift and erosion, during which the Sierras were raised to great heights and fractured, and gold was formed in the fault zones; and will describe how these disturbances turned the rivers from their older north-south courses to their present east-west trend, and enabled them to erode the gold that had been buried by lava in their old channels and unlock the ore which was imprisoned in the quartz seams. The layman sees, however, not the old hand of Nature at work, but a landscape that has been altered radically by the nozzles of hydraulic hoses in the hands of men, making hills and valleys where none existed. St. Anne's Roman Catholic Church at Columbia, built on what was once level ground, now stands at the top of a hill upon gold-bearing gravel that the priests refuse to allow to be washed away, surrounded by a desolation of hydraulically eroded hollows.

But there is no ghost town atmosphere about Sonora, the gem of these southern mines for, like her sister Grass Valley to the north, lumber has given her a second lease of life, and cattle and ranching a more enduring wealth than gold. Here, where in the cool of the evening cowboys from the range head for the shadows of the saloon,

---

[1] Bulmore Mss., testimonial of Hennen Jennings, "Superintendencia de la Compania Minera 'El Callao' ", 26 July 1889.

is the oldest Episcopal Church in California and a fine county museum that displays a wooden cradle brought from Cornwall in 1867 by Mrs. Edwina Carne when she came to join her husband, the engineer at the Soulsby mine. It is also the home of two colourful personalities of the West, Mr. and Mrs. Herbert Hamlin, who publish *The Pony Express*, a magazine that attempts "to rescue for America the old Pioneer Spirit by relating true stories of Famous Frontier Trails". Hamlin hails from South Dakota and graduated from a mining school in Washington, but it was from Cousin Jacks in Utah and Montana, he says, that he really learned the art of mining as he worked underground with them, boarded with them and exchanged for their pasties his "dago red", an Italian claret so rich it had to be "cut" with water. An athlete himself who once raced against that amazing Indian runner, Jim Thorpe, his Cornish hero is Bob Fitzsimmons, the blacksmith from Helston who was the only man ever to win world boxing championships at three weights, the more incredible since he was of so slight a build; and Hamlin attributes his skill to a punch concealed in a left fist that had to travel only six inches to reach its target, the result of twisting pieces of iron on the anvil to give him a wrist as flexible as a universal joint. But his wife's kinship with the Cornish is even closer for she married one of them, Richard Jose from Lanner, a blacksmith, the singing sensation of San Francisco and America's most renowned contra-tenor.

No boy could have faced life with so many handicaps as Richard Jose. His father died in 1878 when he was only nine years old and there were three children younger than himself to be cared for by his widowed mother whose small income had suddenly ceased. An uncle in Virginia City therefore offered him a home and sent the fare. So, incredible as it may seem, with a fourth-class ticket in his pocket and an addressed label sewn to his jacket, he was put on a train at Redruth, his ultimate destination 4,500 miles away. It was January 1879, the very worst time of the year for an Atlantic crossing, as the passengers of the Swedish emigrant ship he boarded at Plymouth discovered; but he apparently boosted their morale by singing English hymns and Cornish carols with such effect that they called him the "Singing Kid". Bewildered and penniless in New York, fed by a Swiss family on the overland train for California, he arrived in Virginia City to find that his uncle had disappeared. For a year somehow he fended for himself in this strange cracked wilderness of desert and canyon, delivering bread to miners and earning just enough to buy a bed in a back room of second-rate boarding houses. Then he tried his luck at Carson City, walking across twenty-five miles of rugged country where even today leather is no match for its rocky surface. There he soon became a favourite with the miners as he entertained them with his songs in their smoke-filled and whisky-perfumed saloons; but the Women's Temperance League objected to this corruption of innocence and had him removed by the sheriff to Reno, where in a miraculous way he stumbled upon a relative, Bill Luke the blacksmith.

Here he worked for several years, pumping the bellows, twisting the hot metal, expanding his lungs for the singer-to-be, and attracting the custom of Theodore Winters, the racehorse baron of Nevada, and the concern of Bishop Whitaker, who arranged for him to have singing lessons. When he was sixteen the chance came to join a minstrel troupe that was touring the mining camps, and from now onwards his future was assured. So sensational was his appearance in San Francisco that within a matter of weeks he was offered an engagement on Broadway by the impresario Lew Dockstader for a fee of 75 dollars a month and billed as the "phenomenal alto". After winning a gold medal for ballad singing at Carnegie Hall, he was given a contract to appear in the musical *The Old Homestead*, and tours in South Africa and South America followed. For thirty years he appeared in vaudeville at Madison Square Garden, then the largest motion picture theatre in the world with a screen 100 ft wide, singing in the wings to his gigantic image and delighting thousands, including that veteran of the Civil War, Sherman, with the songs he made famous in every home: *The Sunshine of Your Smile, Just as the Sun Went Down* and Paul Dresser's *The Blue and the Grey. Goodbye Dolly Gray* he once sang dramatically for Theodore Roosevelt and his Rough Riders as they swept down Broadway bound for the war in Cuba.

But perhaps this lad from Lanner is now best remembered for *Silver Threads amongst the Gold*, a ballad of undying sentimentality that had been written by H. R. Danks for the World Fair at Philadelphia in 1876 and then forgotten until Jose discovered it in a second-hand music shop there. He was so impressed with the simplicity of the words and the melody that he persuaded the Victor Talking Machine Company to allow him to record it and so helped them to a fortune. But, ever mindful of his own struggles, Jose made it his business to search out the composer, found him ill and living in poverty, and induced the publisher to pay him royalties. It was an action typical of the high ethical standards he always set himself. An abstemious man, who never drank liquor or smoked in compliance with a promise to his father as he lay dying, he was a devoted Mason of the Ancient Arabic

Order of the Mystic Shrine at San Francisco; and from their members organised the Joseans, a choir that appeared on parades at the Hollywood Bowl, at State Conventions, and in the streets of the city singing Cornish carols at Christmas-time. So dependable was his reputation with politicians and business men alike that the Governor of California appointed him Deputy Commissioner of the Real Estate Commission to raise the professional standards of brokers and agents, a responsibility he was discharging successfully to the time of his death in 1941. Public figure that he then was in the worlds of music and business, he might well recall the days when he was described by a ballad writer of the Sierras as:

> Youthful minstrel of the Comstock,
> Carson's barefoot ballad boy,
> Who filled saloons with Cornish tunes
> And miners' hearts with leaping joy.

Long before Richard Jose was born, Cornish miners were treading this countryside fragrant with wild lilac, dogwood, pink mountain heather and wild roses, their "mountain misery". In 1851 Ben Soulsby, a sheepherder, discovered the famous lead to the east of Sonora that produced more than 6,500,000 dollars' worth of gold. Tradition and a fine historical marker in the town that bears his name claim that in 1858 as many as 499 hardrock miners were imported from Cornwall to work his mine, though this is somewhat doubtful for the census returns of 1860 reveal nothing like this number, but a mere 129 English miners for the whole of Township No. 1 that included Soulsbyville. What proportion of them were Cornish it is quite impossible to tell, for street directories, so valuable at Grass Valley and New Almaden, are not available. Yet 1860 is the year when the foundations of the Methodist Church were dug; and one of the money-raisers was the Cornishman William Inch. An attractive building of unusual design, being semi-circular in shape, it contains two stained-glass windows, one to William Curnow and another to a Nichols. The Cornish here have not been forgotten in other ways; for instance in 1958 the Tuolumne County Historical Association arranged a symposium on the Cornish contributions to the development of Soulsbyville, its Vice-President, Donald Segestrom, claiming that their skills "were used everywhere much to the profit of mine owners and stockholders" and that they took out of the mines almost 150,000,000 dollars in gold[1].

This would suggest a much larger number than those mentioned in

---

[1] *The Tuolumne Prospector*, 30 October 1958.

the *Great Register* of Tuolumne County for 1873: James Bawden, Joseph Barron, John Bluett, John Curnow, Edwin Carne, John Carne, William Gribble, Henry Gundry, John Hawken, John Hoskins, Thomas Hodge, Sam Hender, John Johns, Vincent Johns, William Moyle, William Nichols, the three Peters brothers, John Richards and William Trengrove. A generation later the colony has changed little, even though in the 1880's mining was "at a painfully low ebb"[1], but there are a few fresh names: Thomas Dunstan "blind in his right eye"; Joseph Henry Mitchell with "a forefinger on his right hand missing"; and a Nichols who is the superintendent of the Draper mine. A few have found their way into print. John F. Bluett crossed the plain from New Almaden with Joseph Peters in 1864, located the abandoned Landers mine, changed its name to Black Oak and opened it to a depth of 1800 ft so that it yielded 3,000,000 dollars in gold. Richard Inch was in charge of the Soulsby mine in 1865 when it made one of its most spectacular runs, its twelve stamps in one week disgorging such a volume of gold that its value was never disclosed[2]. In 1882 Richard Johns was modernizing the same mine where "a very large sum of money has been expended on permanent improvements and everything metalled for the profitable extraction of the ore"[3]. John Lobb, after mining in Pennsylvania, Michigan and Colorado, arrived in Sonora in 1895, married there, started a family in Jamestown nearby, and then prospected in British Columbia and Alaska, his sons rightly complaining that they could never live together as a family. Fred Richards tended a horse whim at the Independent. John Richards, Joseph Ede and Andrew Carkeek are buried in the cemetery on Cabezut Hill. John Rosewarren was trumpet major of the Soulsbyville Silver Cornet Band. The Barron family produced two farmers, a blacksmith and a storeman. Matthew Hodge, after a lifetime of wandering, settled his family here while he made one last bid for fame and fortune in Alaska and then returned to see his son Havilah, once a miner, teamster and carpenter, appointed a judge of Tuolumne County after studying law in his spare time. The judge's death in March 1965 removed a figure whose life was an example of the frontier as a continuing transformation scene, for part of it was endured in one of the most heartless regions of California, the Mojave Desert.

The gateway to Mojave from the south is Barstow, once a historic cross-road for Ute, Pahuti and Mojave Indians, Jesuit missionaries

1 B. F. Alley, *The History of Tuolumne County* (San Francisco: 1882).
2 Edna Bryan Buckbee, *The Saga of Old Tuolumne* (New York: 1935), pp. 310-11.
3 Herbert C. Lang, *A History of Tuolumne County* (San Francisco: 1882), part 3.

from Spain and silver prospectors. With its gesticulating Joshua trees that a Mormon battalion believed always pointed to the promised land of Salt Lake City, the desert could well be a nightmare for the lost prospector of yesterday and the unwary traveller of today. Across its baked and cracked floor, where the only rain is an occasional cloudburst, often so devastating in its effects that deep ditches have to be dug into the only road that crosses it to preserve both road and water, sprawls a mountain range in which was once locked a fabulous hoard of silver. It was prised open by miners who found themselves in one of the most eerie places in the world and in 1881 called their camp Calico because the colouring of the mountain was "as purty as a gal's calico skirt"; and this is an exact description of the rocks, striated and patched as they are in gay shades of red, pink, brown, yellow and ivory. Twice destroyed by fire and ruined when the Government devalued silver, so that its price slumped from 1.3 dollars to 63 cents an ounce, Calico lost all its records, account books, pay rolls and its delightfully named newspaper, *The Calico Print*.

In 1885 there were about 2,000 miners in this timberless, waterless and hopeless region though, since there was little rain or snow, conditions were bearable for most of the year, apart from the merciless winds that strafed the mountain walls, forcing the Cousin Jacks to dynamite holes for shelter, making them cliff-dwellers like the Mojave Indians. It is not likely that many Cornish wandered here, for this was a poor man's bonanza; so compact was the body of the mountain that tunnelling could be done without timbering; the mines were no more than small cavities in the teeth of a slumbering dragon (one mine indeed was called the Dragon); and their names Silver Odessa, Snow Bird, Lone Star, Black Cloud and Josephine, only betrayed the melancholy hopes of those who ached to find a fortune within a fortnight[1]. Here they lived in their holes in the ground, only venturing into Calico on Saturday nights for the gay "Grand Ball" at the Odessa Boarding House, from which the dancers would depart in their barouches, chariots, gigs, buggies and rockaways for a moonlight picnic in Mule Canyon. On the whole Calico was a quiet town and free of brawls, folks making their own music and even attempting a literary society. Quacks and rogues of course abounded, like Dr. Kellog who claimed he could cure cancer painlessly and "without cutting" by means of his "magnetic needle". But death here came insidiously from the lack of fresh water, which had to be hauled from wells near Calico Lake; and

---

[1] *The Calico Print*, 22 February 1885.

many children fell victims to typhoid. On Boot Hill is a little grave covered with rocks and marked by a headboard to a Cornish child which simply states: "1884. Baby Ernest Rowe. Aged two years. Typhoid".

Two other Cornish are known to have been buried here, though their graves have long since disappeared, both children of Matthew Hodge. A native of Blackwater, he was the world's wanderer to the despair of his wife. As soon as he was married in 1861 he carried her off to Australia, mined there for ten years, gave her two children, a son in Victoria and a daughter in New South Wales, and then returned to Cornwall. For the next ten years he was away on his own in the Upper Peninsula of Michigan, having blessed her with another daughter before his departure. In 1881 she obeyed the call to rejoin him, somewhat reluctantly it seems, for within a short time her husband was on the move again; first to the heat of New Mexico to prospect for silver at Lordsburg, where a boy who was to become a judge was born in an adobe cabin filled with armed miners desperately beating off an Apache attack led by Geronimo himself; then to neighbouring Arizona and far-away Montana; and finally to Calico in 1888. Within a few weeks of their arrival the greatest calamity of all overtook them: the son born in Australia and the daughter born in Cornwall both died of typhoid.

Calico broke the hearts of the Hodges so they left the desert for New Almaden, Grass Valley and Soulsbyville, where the more settled life helped them to forget their tragedy. A ranch was bought near Sonora and from there Mrs. Hodge refused to move, her husband spending his last twenty years roaming from camp to camp and from continent to continent: South America, South Africa and even Alaska, only returning "home" to die in 1918.

Sculptured by fire and ice since the beginning of time, chipped and chiselled by thousands of picks and gads, the mountain rises bold and clear; but it holds within its caverns and tunnels the secrets of the Cornishmen who worked it. Accounts of the Cousin Jacks at Calico are rare, so the following is welcome. Herman F. Mellen first met them in 1885 at the Garfield in Odessa Canyon, where they comprised about half of the labour force, "the best miners that the world could boast, having followed the trade, father and son, for centuries". One he remembered was "big" Jack Pascoe, six feet tall, who had been hired with two other contractors to drive 400 ft into the King mine: "Putting in every third shift in the tunnel, Jack raised its roof at every shift by at least one and a half feet to accommodate his great height. When the annoyed contractors protested, his reply was, 'Damme, old son, you have to maken place for my feet' ''. So the irregular roof in the tunnel became known as Jack Pascoe's mark and, when Mellen visited it in 1941, he could tell how many shifts Pascoe had worked[1].

The Mojave Desert is part of that vast region of wilderness which includes Owen's Valley, Panamint Valley and Death Valley. Tomesha or "ground afire", as it was known to the Panamint Indians, exactly describes Death Valley. Even at Christmas-time shimmering white and yellow under a blue sky, 130 miles long and up to 15 miles wide and flanked by the formidable Panamint and Funeral Mountains, it is ageless and timeless, both incredibly attractive and strangely forbidding. Today it is the workshop of the geologist and the naturalist for here is life that began before history was. Even the sand-dunes, smoothed by the winds into crescents and parabolas, have a look of permanence. A macabre beauty haunts its rock formations of a variegated pastel colouring, some stripped and chevroned like an Indian blanket, nowhere obscured by soil or vegetation. Yet nowhere is there a peak, a canyon or a flat to suggest God their Creator; they are all the work of the Devil, the Devil's Golf Course and the Devil's Cornfield looking like Dante's vision of Hell. Death Valley's floor is an alkaline sea, a really dead sea that a relentless sun has sucked almost dry, with a bed of salt more than 1000 ft thick, the boothill of desert schooners and men of metal who worked the "graveyard shift", at "coffin mines" on "tombstone flats".

The most haunting quality of this empty land is its silence, as if all the sounds of the globe had been lured here by winds and hushed for ever in a lunar landscape of steep alluvial cones and alkaline pools. The valley reaches its greatest depth at Badwater, 279.6 ft below sea level where, above a splash of torquoise set in a rim of golden stone, Telescope Peak rises sheer to a height of over 10,000 ft. Under happier circumstances it would surely have merited a name other than the valley of death, but such it was to William Lewis Manly and the party he was trying to extricate in 1849 on the short cut from Wisconsin to California. He most certainly knew the Cornish lead miners there for he recalled that they would not accept paper money and preferred silver dollars or English sovereigns. But there is no evidence that any of them accompanied him on the fateful expedition through Death Valley of which he has recorded: "Thirteen of our party lie unburied on the sands of that terrible valley, and those who lived were saved by the puddles of rain water that had fallen from the small rain clouds that had been forced

1   Herbert F. Mellon, 'Reminiscences of Old Calico', *The Calico Print*, December 1950.

over the great Sierra Mountains in one of the wettest winters ever known. In an ordinary year we should all have died of thirst"[1].

Vernon Tregaskis is a Cornishman who worked his perilous passage through Death Valley and came out alive. Of a line of miners and freighters of Idaho's Silver City who had emigrated to Portland, he was unaware of his Cornish backgound until he was attracted to a supposed bonanza at Quartzburg, near Idaho City where, because of his name, the Cousin Jacks welcomed him as one of their own. Too tall really for mining, his first venture was to Butte City, which he never relished because it was so humid and hot that work was only possible for twenty minutes at a time. Then about 1900 he entrained south for Rhyolite, that fantastic city in the Tonopah Mountains, sometimes described as the last flowering of the mining boom, where men with little money but big hearts hoped to found a city destined to be the capital of southern Nevada. But Tregaskis cherished no illusions about these "rainbows chasers" and their "goldfield fever". Rhyolite, he says, was a stock company's perpetuation of a swindle. So many honest investors were tricked into being persuaded that another Comstock had been discovered that by 1906 almost 10,000 men, women and children were crowded into thin tents and canvas huts, buying water for five dollars a barrel, and in the winter huddling round stoves fired by sage brush because there was no timber. He saw New Yorkers alighting from trains, clutching in their hands brochures of the Rhyolite bonanza that actually depicted steamers on the Amagustro River being loaded with ore; but in fact there were neither steamers on, nor water in the river. So inhuman a trick decided Tregaskis to quit hard-rock mining; and he was lucky to escape with his life, for he was descending a 200 ft shaft in the bucket with a Dutchman when the brakes refused to bite because the engineer had greased the brake drum, either in error or out of malice, and the bucket was halted only six feet from the bottom of the shaft by the resourcefulness of the assistant engineer who threw a wad of cotton waste on to the drum to arrest its progress. As soon as he reached the surface with the unconscious Dutchman, Tregaskis turned his back on the city he knew to be doomed; for it was spawned on the very edge of Death Valley, where water and timber were scarce, transportation costly and the ores of low grade; nor was it likely to attract investment at a time when money was needed to rebuild San Francisco after the earthquake and fire of 1906.

---

[1] William Lewis Manley, *Death Valley in '49*, reprint of a copy lodged by him in Library of Congress, 1894.

It was in August of the same year that Tregaskis became a prospector in Death Valley, looking for silver; it was a decision, he says, that he must have made in ignorance of its harsh nature, for no one in his right mind would ever consider tackling the Valley in the middle of summer. He remembers that what appeared to be in the glare of the sun a vein of silver turned out to be only asbestos. With a pardner and a burro he picked his way through the Funeral Mountains, so called because their light coloured rocks are capped with heavy masses of basalt that gives them the appearance of being fringed with crepe[1], and struggled to Stove Pipe Wells, where the air temperature in the late afternoon was an uncomfortable 135° and the water was brackish and odious. He would not drink and would have perished of thirst but for a supply of tinned tomatoes he had bought in Rhyolite. Another recollection was finding himself crawling with lice after a night's sleep in a bed of leaves that must have been occupied by some other prospector. How to delouse without water became a problem of acute dimensions until he remembered that red ants eat lice. So he stripped to his boots and hat and threw his clothes on to an ant heap, little realising that the ant is a slow eater and that it becomes very cold in the evening at an altitude of 7,000 ft.

Many more like Tregaskis plodded across the deserts of California and climbed up the spine of the Mother Lode. From San José or San Francisco the Cornish took the old trail through the Pacecho Pass and rolling golden hills, studded with live oak and sycamore, and smoothed by the wind and burnished in the sun to give them from a distance the appearance of sand dunes. They crossed a plain of hog-wallow country that canals have now transformed into fields rich with barley, lucerne and rice. They rested at Merced, resplendent with its street palms from Phoenix and white minarets on the court house, from the roof of which rises the figure of Minerva, symbol of California's quickening to statehood in 1850. From Merced they hurried into the foothills and pitched their tents in clumps of blue oaks, digger pines and tarweed. Some lingered in Mariposa, the beautiful land of the butter-fly, and built the Methodist Church: the Tresidders worked the Whitlock mine in 1859; and a Pascoe lent his name to the local historical society's museum that contains a fine reproduction of a miner's cabin. A few, perhaps, found their way to the great forest in the heart of the Sierras down the matchless Briceburg Canyon and through slumbering olive-green mountains, where all is a tangle of wild lilac, lupin, godetia,

---

[1] W. A. Chalfont, *Death Valley* (Stanford: 1951), p. 49.

dogwood, madrones, manzanita and monta rosa beside the plunging waters of the Merced River.

At the head of this canyon is Yosemite, so named by the Americans after its Indians who were "killers of the grizzly bear", though the Indians called it Ahwahnee which means "a deep grassy valley". Its development as a Californian National Park has a special connection with Cornwall. It began in 1899 when David A. Curry, a schoolmaster at Redwood City, started a summer guest camp at the foot of Glacier Point that was to grow into the fabulous Camp Curry of today; their daughter, Mary, married Donald Tresidder, the son of James Treloar-Tresidder from Wenk on the Helford River in Cornwall; and it turned out to be a unique partnership of great benefit to the worlds of business and scholarship. Donald Tresidder, like his Cornish father, was intended for the medical profession, and studied at the universities of Chicago and Stanford until 1914 saw him enlisting in the Army Air Corps as a pilot. When the war was over he returned to his medical studies at Stanford and at the same time worked in the family business at Camp Curry; thus in 1925 he was President of the Yosemite Park and Curry Company and in 1927 a fully qualified medical doctor. By all accounts he was a brilliant administrator, whether in the woods of a National Forest or on the campus of a famous university, for in 1943 he was appointed President of Stanford, an office he still held at the time of his death in 1948. While he is likely to be remembered for a long time to come by the teachers around Palo Alto, most Americans will think of him when they spend a night at Sunrise Camp, one of the highest points of the Sierras at an altitude of 9,460 ft, for this latest of the Curry developments is his wife's memorial to him.

And in startling contrast, almost at the very centre of the valley below that has been cut, carved and shaped by rivers and glaciers through countless ages, is a small cemetery, enclosed by stone walls and cedars, in which are buried a small company of mountaineers, rangers, stage drivers, labourers and Indians. There is one miner and his name was Woolcock, a Cornishman who probably came from Mariposa and was employed on the Coulterville and Yosemite wagon road. One Sunday in June 1874 he fell from a log at a crossing on Cascade Creek, broke his neck and was drowned. Aged sixty, "native of England" and dying alone amidst God's grandeur, he surely has his memorial, unique among pioneers.

IV

THE COPPER AND IRON WILDERNESS OF UPPER MICHIGAN

THOUGH most Cousin Jacks fell under the hypnotic influence of California and the prospects of quick fortunes, many were content to mine the metal they knew best in the northern parts of Michigan, where as early as 1849 they began chipping away the metal which was the foundation of their own kingdom in Cornwall. The true copper country of Michigan lies at its most northerly tip on the Keweenaw Peninsula, a fir-covered craggy strip about fifteen miles wide and fifty miles long that juts into Lake Superior as Cornwall does into the Atlantic.

Even to the most casual observer the similiarities between the Keweenaw and the Cornish peninsulas are most striking. They are almost separated from their mainlands by stretches of water, the Portage isolating Keweenaw and the river Tamar dividing Cornwall from England; and they are both at the extreme ends of long lines of rail communications from their capitals of Chicago and London. The railroad from Chicago to Hancock that carries the Copper Country Limited appears to run no more smoothly than it did half a century ago, though the traveller is richly compensated by a Hiawatha land of mist-curtained lakes and deep forests where the canoe is seen more often than the cabin-cruiser. It serves a population about the same size as that of Cornwall, say about 350,000, though by comparison Cornwall must seem overcrowded; for the Upper Peninsula, which includes Keweenaw and important iron-fields as well, stretches from Ironwood in the west to Sault Ste. Marie in the east, a distance of 300 miles, and thus is ten times bigger than Cornwall. But in both countries population is both declining and ageing as their metals pinch out. Beyond Hancock and Houghton is a Cornish wasteland of decapitated chimney stacks, roofless engine houses, derelict company houses and offices; of the rusting machinery of the once mighty Quincy mine; and, saddest sight of all, of the windowless schoolhouse, for the young men and women have left to rear their children elsewhere. Each has its far-famed School of Mines, one at Houghton and the other at Camborne; neither can survive except in some new relationship with other branches of technology. Farming can be no substitute for mining for the soil in both of these lands of metal is acid and thin, so they both look to developing a tourist industry; and for Michigan the outlook is indeed brighter than in Cornwall. The White Pine Copper Company is still most productive; old tailings are being worked; and it is still an undiscovered wilderness where fish choke its streams and deer swarm the Huron and Mohican forests of aspen and laurel, a cooler for those who want to escape the humidity of the eastern summer, and in winter a paradise for ski-jumpers and hunters.

Here, where the ground was once heavy with copper and iron and is now sullen with deserted pits and hoists, the Cornish have their adopted home. At Iron River Lionel Sleeman, whose father was the superintendent of the Forbes iron mine and hailed from Liskeard, graduated as a teacher from Albion College, that fine training ground for Methodist ministers, and then secured his future by a career in insurance. Old Mrs. Godfrey, born in Calumet of parents from Tuckingmill, remembers her grandfather "who died at sea, was sewn in canvas and thrown to the whales". William Johns from St. Neot is a miner-turned-embalmer who paid for his professional training in Philadelphia by odd-job decorating, knowing that his newly-acquired skill could be turned to good account in a climate where burying is only possible from May to November. John Grenfell from St. Just thunders from the pulpit. Percy Treloar recalls his mining days at South Crofty and Tincroft, one of the oldest of Cornish mines. Leo Collins grows nostalgic for Trewartha Farm, near Mithian, where he was born in 1898, and for Wheal Kitty where he mined. The frail octogenarian Percy Smith from Pensilva in 1961 is the oldest surviving veteran of the Spanish-American War in Ironwood. Noah Warren recalls the time when he was a student at the School of Science in Penzance. And everywhere the Cornish pasty reigns supreme, relic of the days when it could be reheated in the mine on a Cornish stove, that is, two candles beneath a shovel.

Michigan is arrowed by immense distances and the long drive northwards from Ironwood to Hancock, with the waters of Lake Superior to the west, is not unlike the much-travelled main road down the spine of Cornwall. Motoring along endless ribbons of pavement by boot-shaped Lake Gogebic, through mile upon mile of forest to Bruce Crossing and thence to Mass, named after a colossal chunk of pure copper found there, you finally arrive at the "villages" of Hancock and Houghton, where the Cornish family trees begin to spread their branches. One that originates with the Wilcoxes and the Edwardses from Penzance in 1875 discloses about eighty of Cornish descent in the Houghton district alone. Here is the gateway to the Keweenaw Peninsula where the Methodist Church is still predominantly Cornish.

Here is a Dakota Heights Motel that recalls a farm called Dakota on the moors between St. Just and Penzance, built by a Cornish miner from Michigan with his own hands from the outcropping granite, Will Green of Morvah, who in 1905 sailed on the *Philadelphia* from Southampton to New York for £9, mined at Trimountain for 60 dollars a month, and then returned to his croft to name his farm after a Dakota girl.

The Keweenaw Peninsula, almost an island of copper ore, where the settlements are villages rather than cities, has attracted the attention of many writers because of its Cornish character. James Fisher, the American historian of the region, generously observes of the Cornish:[1]

Though in most cases not possessed of any great degree of book-learning, the natural shrewdness and almost instinctive knowledge in mining affairs, inherited from generations of those who had preceded them in the same calling, made the Cornish leaders in their work and in their community ... No mining community in the world can boast of a more loyal group of former employees than Central in Keweenaw County. They were loyal to the Company and to the community and though the location is now practically deserted, no mining having been done on the property for thirty years, the annual Central union is looked forward to and participated in by all the old time residents.

This Central mine is also the location for a novel, *The Long Winter Ends*, by the Cornish-born Newton Thomas[2]; and no mine could have been more suitable for it was so Cornish that it was known as Keweenaw's Duchy of Cornwall to which Cousin Jacks "headed like homing-pigeons"[3].

The story begins in east Cornwall with the closing of the mines around Kit Hill and the attempts to raise passage money to Michigan from relatives, tradesmen and even the doctor. Its hero is Jim Collins, singer of the hymn *Diadem* in mine and chapel, who is urged to leave Cornwall for ever by an old captain who, having sampled the new America in Michigan, regretted his return to the old degrading life in Cornwall: "For the man who is not afraid of work, the man who invests himself, the intelligent man, it is a Land of Promise". Four weeks later Collins and four companions are standing on the wooden platform of the small railroad depot at L'Anse, bewildered by the French and Indian place-names, and shivering in the high wind blowing

---

[1] James Fisher, 'Michigan's Cornish People', *The Michigan History Magazine* (Vol. 29, 1945), pp. 377-85.
[2] Published by MacMillan, New York, 1941.
[3] Angus Murdoch, *Boom Copper* (New York: The MacMillan Company 1943), pp. 201-2.

across Lake Superior. Here at the edge "villages sat in a welter of logs, of sawdust, of bark and ships", while "across the bay, three miles away, forest ranged north and south and west as far as the eye could see, its furthermost edge meeting the sky". From L'Anse they embark on a steamer for Hancock, thence by train to Calumet, and finally by coach, fitted with runners and its floor strewn with straw against the penetrating cold, to the Central mine.

The novel is primarily a tribute to the skills and the adaptability of the Cornish in the last twenty years of the nineteenth century. Its heroes at first have no other ambition than to make money and return home, but gradually they come to terms with the new frontier and send home for their wives and families: "Necessity do work wonders. 'E will change Englishmen into Hamericans". Other nationalities acknowledge their excellence: "Given a hammer and a few drills, these men would make passage through anything that steel would dent and powder break". They are praised for the keeping of the Sabbath: "You'd think Cornwall was a chip off the 'oly Land an' all the Cousin Jacks pious". But the reader is left in no doubt about their influence: "The flavour of Cornwall will last in the Peninsula a long w'ile hafter the Cornishman be extinc' ".

Yet they do not escape criticism. They are too clannish even to attend the funeral of a Finnish miner. The lay preacher sees "the steps to the pulpit as a ladder leading out of the mines", and some even manage to "get on the plan" without being able to read, for "the usual phrases of the pulpit did the trick". Worst of all, they are seen as feckless and transitory birds of passage: "You sleep in beds that can't cool between shifts and say nothing—worse, do nothing . . . You live with your bags packed for a hurried getaway. You own nothing and care for nothing . . . You have no responsibility and no interest . . . You raise no garden, plant no flowers, maintain nothing, because you are not going to stay". It is said that there is a wretched sameness about them that amounts almost to a disease; they are all Methodists and show no curiosity in other faiths; they all sin against their beliefs; and they know nothing about politics. "All they have is the mine, the town, their hungers and memories."

With so many Cornish entrenched in the Keweenaw Peninsula in the 1880's, the legend has persisted that they must have been there even since 1771 when a London company under the direction of Alexander Henry began operations in a clay bank on the Ontonogon River. Though this is doubtful, it is fairly certain that in the summer of 1844 some twenty or so were working for the Lake Superior Copper

Company in the vicinity of Eagle River, and that they were the first real miners to reach this district[1]. They may have come from the lead region of south-west Wisconsin or from the Bruce copper mine near Thesalon in Ontario (where today headstones proclaim the Trevillions, the Tregonnings, the Tretheweys, the Tremaines and the Trelawnes, some of them from Newquay and all victims of typhoid); and the journey could not have been too difficult for the fur companies were operating their schooners on the lakes. Among them were Job Masters and his wife Jane, perhaps the very first Cornish folk in Eagle River[2]; and two years later they were joined by Captain John Hoar, whose mining experience had been gained while working for a London company in Ireland and Germany. In July 1846 he landed at Copper Harbour where "there were only three families on the whole Point (Keweenaw Point) and no accommodation for strangers, so he was obliged to return to the boat to sleep until some shelter could be prepared". He was an employee of the Boston Copper Mining Company and his mission was to begin mining operations on their location of four square miles which was to become the Boston mine. More Cornish arrived within the next two years, for we know that in 1850 Joseph Paull from Wisconsin, employed by the same company, married a Cornish girl, and that "this wedding was about the earliest in this region"[3].

Conditions here were extremely difficult and hardly to be tolerated; navigation on the lakes was possible for only five months because of freezing and the winter storms. The few harbours were blocked by sand-bars so that passengers and freight had to be transferred to lighters; and the country was usually a howling wilderness of swamp and forest, impenetrable between November and May except by the most experienced trappers equipped with dogs and sleighs. Only the Cornish miner from Wisconsin could have survived so savage an environment for he was already conditioned to snow and ice and to a form of mining that differed little from that in Keweenaw, where the lodes were copper masses of almost pure metal which only had to be cut up underground and shipped directly to the smelters. It is not surprising therefore to find that the first mining methods in Keweenaw are Cornish, brought from Wisconsin, and belonging to the Cornwall of an earlier age before the deep mine required a beam engine to free

---

[1] *A History of the Upper Peninsula of Michigan* (Chicago: The Western Historical Company: 1883), p. 276.
[2] J. B. Martin, *Call It North Country* (New York: 1944), p. 80.
[3] *A History of the Upper Peninsula of Michigan*, pp. 281, 317.

it of water. Drilling is by hand with sledges; blasting is by black gun-powder; haulage is by the Cornish "kibble" powered by a Cornish horse-whim; the miners ascend and descend ladders; the stamps are of the standard Cornish pattern of wooden stems supporting iron heads that weigh 200 lbs and drop on the rock twenty times a minute[1]. And for the first time in Michigan is heard the new language of an old technology; a fine lode is a "brave keenly lode"; waste rock is "deads"; and water that seeps back into the mine is "mad water"[2].

One of the first mines to operate on the Peninsula was the Cliff. Manned in 1849 almost entirely by Cornish miners who had survived the slump of 1847, caused by a weakening of world prices and the prohibitive costs of shipping the ore to Boston (20 dollars a ton, com-pared with 15 from Chile and 6 from Cuba)[3], it was modernized by the Cornishman Joseph Rawlins from the Canadian Bruce mine, who in 1848 installed the first man-engine[4]. Two years later the company built its first steam winding engine, straightened and widened the shafts, substituted skips for kibbles and improved on the small Cornish "pestle" stamps by using iron shafts and driving them with more powerful steam engines[5].

But whether the Cornish would stay depended entirely on an improve-ment in their living conditions for in 1859 it was said that they were only saving up for the great day when they could return to Cornwall with their pockets full of gold[6]. It was still a country fit only for a single man, where rapacious landladies in primitive boarding houses charged him half of his monthly wage of 34 dollars[7]. As in California, brawls between the Cornish and the Irish were frequent; in 1857 the body of John Terrell was almost bisected with an axe outside Dan Ryan's saloon[8]. Communities were small, twenty or so miners and a handful of women and children, huddling together in log bunk-houses for warmth and knowing the meaning of hunger. Conditions underground were unpleasant rather than hazardous; rock had to be removed by

1 W. B. Gates, *Michigan Copper and Boston Dollars* (Harvard: 1951), pp. 4, 5.
2 James P. Jopling, 'Cornish Miners of the Upper Peninsula', *The Michigan History Magazine* (Vol. XII, 1928), pp. 559-60.
3 Gates, pp. 3, 6, 8.
4 *A History of the Upper Peninsula of Michigan*, p. 285.
5 Gates, pp. 25-6.
6 Charles L. Fleischman, *The Portage Mine on Keweenaw Point, Lake Superior, a Report of an Examination of the Mine made during the summer of 1859* (New York: 1859), pp. 21-2.
7 Gates, p. 100.
8 James J. Jamieson, 'The Copper Rush of the 50's', *The Michigan History Maga-zine* (Vol. XIX, 1935), p. 383.

hand, ventilation was poor and water froze in the shafts. Cave-ins were unusual, though at Copper Falls near Eagle Harbour, where the hanging walls were very unstable, seven men were buried alive and "before their bodies could be recovered, they were so badly eaten by rats as to be almost unrecognizable"[1]. And at Eagle River one of the first casualties, as a tumbled headstone tells, was Absalom Bennet, killed at the Cliff in 1859 when he was only twenty-six.

Keweenaw became a permanent settlement when the Cliff took on the rôle of the power house of the Union cause. Management encouraged their Cornish employees to rent or buy land, assisted them to build houses for their families, donated money for churches and schools, and even imported the Cornish idea of the "bal surgeon", every married miner paying a dollar a month (single miners 50 cents), in return for which the company doctor serviced the entire family. Most important of all, wages at the Cliff soared to 65 dollars a month[2].

A picture of what life was like for the Cornish at the Cliff during the Civil War has been preserved through the chance discovery in a Wisconsin book store of a journal kept by a teacher at the company school[3]. It consists of two bound volumes covering the period January 1863 to August 1864; and its author was Henri A. Hobart, a young man from the Lake Champlain region and of middle-class background whose father in the summer of 1864 was nominated Senator for New York State. What motives directed him to this outpost of civilisation are not clear, unless it was to escape the army, but to begin with he seems to have been blessed with an abundance of enthusiasm to bestow on the 150 Cornish children "sufficient instruction and of such a nature as will fit them to occupy honourable positions in life". Yet within two years his fire has been extinguished and he returns to Vermont, beaten by the "most God-forsaken place" in the United States and bitter with the Cornish on account of their habits:

"The Fourth was celebrated in this place in the Cornish way. There must be no change—whatever was done by preceding generations must be done today . . . It is the most disagreeable sight I ever saw. They are like hogs in every sense of the word . . . I am disgusted with the want of public spirit in the place. A crowd of whisky soaked Beer Bellys are the blue-eyed set of Cornish."

[1] *A History of the Upper Peninsula of Michigan,* p. 342.
[2] Gates, pp. 100-4.
[3] My attention was drawn to this journal by the finder, Mrs. Elleine Stones of Albuquerque, who at the time was head of the Historical Department of the Detroit Public Library. The journal is now in the Burton Historical Collection there.

Though he may not be a reliable witness because he can never accustom himself to Cornish ways, yet his observations are an admission of the depressing nature of the environment of Keweenaw. By the end of April they were "in a very starving condition and deprived of meat and almost everything else except bread". It is no wonder then that when fresh supplies arrive over the Lakes in May most of the Cornish celebrate the end of winter with "a drop of Beer and Paddy's Eye-Water to 'taper off on'". Some, however, it is said, become "raving mad under the influence of Gin"; while others "abuse their wives and kick them and beat them, pursuing them from place to place until they would hide in some house until the man was no longer drunk". As a class of men the Cornish seemed feckless and irresponsible, forever "shifting places" which Hobart considered was "not the way to become properous". But, on the other hand, no one could be expected to withstand the siege conditions of a Keweenaw winter without it leaving some marked effects on his habits; and this he never really understood. So, unable to adapt himself to these severe frontier conditions, he vents his spleen on Cornish standards of living, never understanding that they were common to all mining camps:

I find little enjoyment. The style of living I should always dislike. It is so seldom that I see a potato on the table that I have almost forgotten their taste. My breakfast is dry wheat bread and water with a little milk in it. The butter is so filthy that I can hardly get along with it on the table much less to eat it . . . I could never enjoy Cornish living.

His sensitive stomach retches at the sight of the women trying to be neat and orderly about their households and yet having to be careful with water. When he suffers from a bilious attack, he says: "I could eat nothing but what I did get was but a dish of soup or the broth that the beef, cabbage and onions are boiled in ($\frac{1}{2}$ greese, bread and onions in it) and perhaps the onions are sliced raw". His jaundiced eye therefore registers a protest against bread being "moulded on a board that looks like a stable window"; against bread being "taken out on the floor when taken out of the dishes in which it is baked"; against dirty dishes being "wiped with the towel they use"; and even against the young Cornish men and girls. All the young men are full of "self conceit and pride and affectation", and as "ignorant as jackasses". And as for the girls:

It is true there are young girls and Cornish girls, but I cannot appreciate their excellent pasty qualities. When I hear a young lady of 180 lbs saying, Now here, he ain't good for noffing for such a brave one as she, Thee art a nice man etc. and this is not an uncommon display of the mental powers of some, I am sick.

Yet to his credit, though he distastes "Cornish twaddle and nonsense", he admits that it is "a shame they cannot read or write", and opens a night school for the parents of his scholars after coming off a ten-hour shift. His criticisms also slacken when he realises that undernourishment and lack of public hygiene are to blame for the deaths of some of his pupils who easily succumb to scarlet fever and typhoid.

Background as well as temperament accounts for some of this sourness. For instance, since he has been brought up as an Episcopalian, he detests the Methodists, even though he has friends among them. He is on the committee to arrange the Fourth of July celebrations for 1864 alongside the Cousin Jacks Abram Trewartha, Thomas S. Williams and William Osborne, who invite him to join their choir. But he refuses since he dislikes John Penberthy who thinks himself the "starpiece"; and urges that Johnson, the Episcopal minister, be asked to the celebrations. But "at such a suggestion Mr. Williams was exceedingly put out and would resist the thing to the bitter end (for) nothing but a methodist can celebrate the 4th here". Since their jollifications on this occasion are to coincide with those to mark the defeat of Robert E. Lee, he cannot resist a few strictures on the attitude of the Cornish to the Civil War:

Cornishmen care nothing about it. They would do almost anything rather than go and fight for the country. Those who enlisted from here after getting their advance pay deserted in Detroit, fleeing to Canada.

This indeed may well have been true though Hobart need not have been so indignant, for many Yankees paid substitutes to take their place in the ranks, while Hobert himself is not entirely above suspicion as a draft dodger. For instance, when he learns on 10 March 1863 that Congress has passed an act conscripting all men between the ages of eighteen and thirty-five, he often wishes that he had "a hand in crushing the rebellion", but he shows no haste to throw away his chalks and enlist.

Only an occasional American is to be seen at the Cliff, which accounts for Hobart being so ill at ease among Cornish, English, Irish, Scots, Welsh and Germans, all of whom look upon water "as only fit for washing copper". When elections are held for town officers, he is disgusted that none of the electors is interested in "the great question that is now agitating our country causing the expenditure of treasure and the sacrificing of thousands of precious lives", but only in fights and betting, bragging and dancing. Rivalry between the candidates from the Cliff and Eagle River is finally decided, not on a democratic vote, but by the Cornish returning officer, "Squire" Vivian who "looks solemn, turn over the law book, fills his pipe and finally decides in favour of the Cliff". Captain Vivian is agent for the Humboldt mine, magistrate and registrar of births, marriages and deaths, and "was called the Governor on account of his haughty appearance". In January 1863 he is recorded as solemnizing the marriage of John Penberthy to Mary Edwards to the accompaniment of a Cornish rattle of old pans and sticks, "practised only by a low class of rowdies", and in July of the following year himself is married to Mary Tresider, sister to John Tresider, "the Goliath of Lake Superior".

Hobart never quite rose to appreciating Cornish wit and wisdom and, when he hears that John Penberthy intends to enrol at Michigan University, his professional jealousy is at once alerted. He admits that he has a fine command of language, possesses a remarkable memory, reads much and remembers it but, he nastily adds, "a thorough scholar who has common-sense will never be continually telling others the principles of Philosophy, Chemistry, Theology and Geology as this Cornishman does who has studied nothing thoroughly but has read a little of everything". Neither has he a kind word to say for the local preachers:

Quite a number of new men have come up this Spring, tho' most of them are old sods from Cornwall. There is one thing worthy of notice with these young Cornishmen—all 'local preachers'. Nothing that I can discover goes to show that they have any education. They can exhort and groan and make a noise but there is nothing refining about them. I listened to one last Saturday evening who went to the pulpit with his hat on—violent gestures, great noise, Cornish eloquence.

A sophisticated and insufferable prig who expected refinements in a mining camp, he pampers himself with visits to Captain Paul's house, "a fine place for civilized persons to live in" compared with the log cabins of the miners, and without the slightest trace of humour adds: "The Captain has made his money out of a small store in selling Goods and Double refined Licquors".

The saleable commodities of the Cousin Jacks, of course, were the labours of their hands and the price was all too often death or mutilation. James Trethewey, one of his adult students, is still on crutches a year after both his legs have been crushed. Samuel James narrowly escapes death when a ton of rock falls on him. William Osborne dies of the "quick consumption", accelerated by walking to the North Cliff mine, getting overheated and then working in a cold drift; his distracted wife faints twice, once in church and again at the grave-side. Samuel Barnes "who was so very seriously hurt in the mine is able to walk around but some of his family are in a very pitiable condition". It is to Hobart's credit that he eventually followed the men "below grass" to see for himself how the parents of his pupils sweated to pay for their schooling and his salary. With Captain Josiah Halls as his guide he makes his descent in January, when the thermometer is well below zero, to the 110 fathom level down ladders slippery with ice, some vertical and some even "inclining under", his only light a candle stuck into a lump of clay on the top of his hard cap[1].

After the long climb down and then up, Hobart concludes that the life of the Cousin Jack is not for him; and he is genuinely astonished the "miners are willing to run through all the dangers attending mining rather than work in the light of the sun". For the first time he understands why they drink and smoke after the shift in levels full of the smoke of blasting powder: "There is no circulation of air, only as it is driven by hand. The miners work in this smoke and of course their lungs are filled with dust. Smoking causes them to spit and throw off this foul matter in the throat and lungs". Seeing them at their work he comes to know them better: Captain Harry George who loans his parlour to train the children for a spelling bee; A. Stoddart, the chief blacksmith; C. Barkel and James Rapson, his companions on many an expedition to the mouth of the Gratiot River in search of agates; Captain Jennings, "quite an old man (who) will climb the ladders with ease"; Captain Sam Bennetts who escorted him underground a second time and afterwards feasted him on "three cases of oysters"; and William Semens and Ed. Richards who were digging a drift towards a "beautiful show of copper at the 120 ft level". All amazed him with their troglodyte agility and the heaven of singing they created around them:

The men wear pants and a sack coat with thick flannel underclothes. The pants and coat are made of bagging and are the colour of copper after being worn once. Thick flannels must be worn as every man comes up very wet from the dripping water. Suspenders (braces) are never worn but a belt about the hips so as to give the arms play in ascending the ladders. The hat is made of wool and resin and as hard as a rock and there is about three inches wide a lump of clay—it holds the candles in front and three or four

---

[1] In Cornwall that clay was near at hand, but at the Cliff it had to be brought from the "Sault".

candles hang on a button of the coat. Thus equipped the miner places his drills in the bucket to be sent down and, with a good crib in his pocket, which generally consists of a Cornish pasty, and plenty of fuse, he is ready to descend . . . How fast the miner ascends and descends on these ladders, singing all the time. It sounds fine through the drifts.

Already a new mine, the Central, was beginning to boom under the stimulus of government orders for brass buttons, copper canteens and bronze cannon. In September 1862 its shares were selling at 4 dollars but by the following April they had soared to 30 dollars and were expected to reach double that figure before the end of the year. Some of the Cousin Jacks were investing their savings in the Central, but the majority, like Hobart, regretted being short of capital: "If I had brought up two thousand dollars last fall, I should be a rich man now". Its appearance was timely for the miners had been drifting away from the Peninsula in pursuit of higher wages elsewhere, some to Montana and other to the iron mines near Marquette; a few took to labouring on the railroads because they could no longer tolerate working at depth down ice-covered ladders; and for many there was the the pull of silver and gold in the West. Hobart observes in June 1863: "There is a great change going on among the miners here. A large number of men are going away . . . Some go below (i.e. to the Portage), some to California, some to Pike's Peak and others to South America. There is a great deal of talk in relation to making a fortune in the gold regions of Pike's Peak". William Trevarthan reported that at the Portage "all the best families in the place with one or two exceptions have gone away". And Hobart seriously considered joining the exodus since "my friend Mr. Trewartha . . . thinks of going to California".

But as fast as the Michigan Cornish departed to risk their necks in the holes of Colorado and Nevada and along the bright seams of the Mother Lode in sun-baked California, their places were taken by others from home. They came in spite of the winter menace of howling winds from the northern arctic wastes that drifted the snow, hardened the ground and piled the sand at the entrace to Eagle River's tiny harbour; and blew the relief steamer on to the reefs, and reduced the communities to a starving garrison sickened by butter that stank and was only fit for making "soap grease". The Cornish adapted themselves well to this climate, the very opposite of English weather: "The people prepare themselves against the severe weather very well; most of them wear shoe-packs and German socks, very comfortable things". And in the winter evenings, "the ladies of the place may be seen out sleigh-riding down hill . . . without regard to age". On the whole Keweenaw was healthy enough, "a great place for a person of slender constitution", for its air was pure and bracing and sickness almost unknown, compared with the Portage where, in the summer, outbreaks of typhoid were frequent, due to the bad water.

One of the most successful Cornish entrepreneurs of the region was Richard Heath Rickard[1]. Born in 1827 at St. Day, he was the eldest of the twelve children of Hannah Heath and Richard Rickard, a mine captain at Wheal Busy, at that time the world's greatest copper mine. The slump of the 1840's forced the family to Ireland, where their father managed a lead mine, until widespread famine drove them back again to England. But by then R.H.R., as he was affectionately known, was already on his way with the Irish to work other mines, Ellenville in New York State and Perkiomen in Pennsylvania, until the summer of 1850 when he arrived on the Keweenaw Peninsula. It is probable that he had already read the government report of 1846 on its incomparable resources, and envisaged the mighty mines that would rise out of its metallic soil; the Quincy, the Pewabic, the Franklin, the Osceola, the Ahmuk, the Tamerack, the Columbian and Wheal Kate (Cornish by name and Cornish by ownership, for it belonged to Richard Edwards, a captain at the Albion), one of the most prolific of the "free copper" mines, where Richard rustled his first job. They went into partnership together, taught newcomers their Cornish techniques, and boldly threw themselves into the rich stream of investment that flowed from Eagle River, down to the new townships of Houghton and Hancock, and east to the banks of Detroit, Boston, Philadelphia and New York. Shuttling back and forth between Lake Superior and the Atlantic water-front, Rickard had so identified himself with his new Cornwall that by 1854 he had become an American citizen. He never quite forgot the old, however, for a year later in New York he married the daughter of Stephen Drew Darke, a poor pilchard fisherman from St. Columb who disconsolately dragged his nets around Newquay for the last time and roughed his passage to New York on the sailing vessel, the *Robert Kelly*.

His flair for business knew no bounds. He was not only a salaried agent for many companies, but invested in many ventures himself, perhaps even buying out the Albion, and pre-empting land near the Cliff in association with some of the most renowned men of that area like Joseph Gay, John Stanton and Horatio Bigelow. This alliance

---

[1] See Donald S. Rickard, *Blessed Shall be Thy Basket and Thy Store* (Exeter, New Hampshire: 1960).

led to the formation of the Copper Range Company and its exploitation of the Seneca, the Ahmuk and the Mohawk, into which Rickard drafted so many Cornishmen from St. Ives that part of it was known as Digey Corner. Since "he had a finger in every mining pie", in 1858 he consolidated his extensive enterprises by opening an office in New York. Here he corresponded with another successful Cousin Jack from Chacewater, Sam Hodge, who later set up his own factory in Detroit for the production of pumps and stamps. It was here that he would interview Cornish miners on their way to the Portage; and from here he would make his annual summer visitations to the Lakes. Only once was he seriously mistaken, when he missed the floating of the great Calumet and Hecla Company. In his early days at Eagle River he knew the engineer Ed. Hulbert who, while laying out the military road between the Portage and the Cliff, had discovered the famous "conglomerate lode" that proved so hard that no stamp could pulverize it[1]. His old friend of Albion days, Richard Edwards, offered to sink 30,000 dollars in the development of the lode if Rickard would do the same. But he hesitated too long and so saw, to his everlasting mortification, the Calumet riding home on Boston rather than Cornish dollars. His one consolation was, however, that he owned shares in the Tamarack, which was part of the conglomerate. Yet he would have been wise to listen to the advice of Edwards for "probably no man in the Upper Peninsula had a better conception of or more correctly estimated the wonderful resources of this region"[2]. He had arrived a year earlier than Rickard to manage the copper claims of the Albion Mining Company, near Eagle River, and he was then forty years old and ready to quit mining for good. But he stayed with the company for six years, all the time investing in real estate, mineral land and mining stock, so that when he died in 1868 he was a man of considerable wealth, owning 20,000 acres of land and 2,100 shares in the Calumet and 300 in the Hecla mines.

Both Rickard and Edwards died in the knowledge that the whole future of their Peninsula lay with the massive Calumet and Hecla Company, backed by Boston dollars, and the thousands of Cousin Jacks who streamed there from the 1870's onwards. By then there was every inducement to attract them, for wages were high: an average of 65 to 80 dollars a month with the possibility of even 100 dollars. The

---

1 An article in *The Detroit Evening News* of 16 October 1899 attributes the discovery, not to Hulbert, but to a Cornishman, Richard Tregaskis.
2 *A History of the Upper Peninsula of Michigan*, p. 279.

mines were generally dry, clean, roomy and well-ventilated, and the deepest provided with a man-engine. The hours of labour were longer than in Cornwall but less exhausting, for the miner "has his work at his door . . . and a few minutes walk only is necessary to reach his place of labour". Compared with conditions in Cornwall, it was reckoned that "creature comforts here are more abundant, everybody lives well and every facility is afforded to those who have a taste for drinking". Though the length and coldness of Lake Superior winters were objectionable, there was the distinct advantage that "there is no mud in winter, nor does it everlastingly rain, as in Cornwall"[1].

An added attraction was the benevolent paternalism of the Company that ensured, through its social and industrial organization, a stability that was completely lacking in Cornwall. The Company made houses available at a nominal rent of a dollar a room per month, provided a medical service, and in 1877 established an Employees' Aid Society. This was financed by equal contributions from men and management, and guaranteed sick benefits of 25 dollars a month, 500 dollars at death and 300 dollars for permanent disability. It operated stores and boarding houses and farmed them out at controlled prices. It built churches and schools, where the teaching was of the most advanced Froebel type, and at Calumet spent 50,000 dollars on a library of 16,000 books. Individualists in the fullness of time would complain about the sickening ubiquity of the company stamp on everything they touched: the hymn books, the water pipes and the red paint that was daubed on every pine-boarded cottage. They would protest against the moral prohibitions that circumscribed their lives: no saloons, instant dismissal for drinking on the job, and deductions of pay for breaches of discipline[2]. Yet without this form of social security it would have been impossible to recruit and maintain the regular supply of first-class labour that was essential if the Peninsula was to become a permanent settlement. On the other hand, the Cousin Jacks were quick to demonstrate that it was their own system of contract working that had stabilized the economics of mining. As at New Almaden, it was so deeply ingrained that almost one employee in six was a contractor bargaining with the mine captains as to the price per foot or yard for a specific job.

Yet it carried its own peculiar problems. The Cornish, shrewdly aware that the system offered them exceptional rewards for their skills, naturally contended that jobs were more difficult than they actually were. They were canny enough to know that the Company could gain

---

[1] *The Engineering and Mining Journal* (Vol. XVI, 1873), pp. 107-8.
[2] Gates, pp. 104, 109, 112.

from the self-imposed competition among themselves, controlling the level of average payments by skilful supervision of the contracts, by keeping their duration short and by negotiating them in accordance with the rates of pay current in the district. Inevitably, when the battle of wits was joined, the Cornish found themselves with a shrinking wage packet. And their complaints were of no avail for the Company retorted, quite justifiably, that they enjoyed an almost free welfare system. But Bostonians then discovered that Cornish individualism was streaked with Celtic obstinacy, and in 1872 faced their first strike, when the Cousin Jacks downed tools for a wage increase and an eight-hour day; and the stoppage at the Portage and in Calumet was so severe that infantry had to be called in from Detroit[1]. *The Portage Lake Mining Gazette* of 27 June was horrified, as it foresaw the onset of the apparition of Trade Unionism. "Nothing", it observed, "more thoroughly un-American, in practice or in principle, can well be conceived". Few Cousin Jacks would have denied this for they too disliked Union agitators, yet neither did they wish to be crushed between management and organized labour. Temperamentally and historically they were unsuited for such a struggle, loathing both the Union organizer and the monopolist company. Somewhat naïvely perhaps, they really believed they were working for no one but themselves. For instance, no Cousin Jack would ever stoop to pick up a tool from the ground in the presence of one of his own Cornish shift bosses in case he gave the impression that he was working for a superior. The mining company therefore had to be careful in its management of the sensitive Cornish, as Angus Murdoch in *Boom Copper* has noticed:[2]

Sweetness and Light did not reign universal in the middle of the Keweenaw Thumb. There were individuals with a holy hate for the company's domination. Cornish shift-bosses often threw up steady jobs under the aegis of Calumet and Hecla for precarious employment at one of the prospect mines some new company was always developing.

One such enterprise of fantastic proportions was to mine a 90 ft-square chunk of rock, Silver Islet, that projected out of Lake Superior off Thunder Cape. The solution of the incredible problem of extracting the almost pure silver ore, which assayed at over 10,000 dollars a ton (the very best on the Comstock was only 3,000), was entrusted to a task-force of thirty-four Cornishmen, Norwegians, Finns, Italians, Irishmen and Hungarians. Three times between 1871 and 1884 General

---

1 Gates, pp. 100, 113.
2 Quoted Gates, p. 93.

Winter tried to smash the enterprise, finally succeeding with a mountainous wave that dwarfed to insignificance all earlier attempts by storm and ice[1]. For other Cousin Jacks, however, the desolation of Nevada, the brutalizing bluffs of Butte and the leaden heat of Calico and the Mojave Desert were preferable to the domination of the efficient Michigan companies, however paternal they might be. And thither many went when in the 1880's Michigan copper was challenged by first Montana and then Arizona.

The Halls from Redruth, who mined in north Michigan, seem to have tasted the ashes of poverty and disaster as long as they could remember. Hugh Hall's father was only eight years old when his father was killed in the mine, and ten when his mother died, so he and his six brothers and sisters were cared for by an aunt, who had no option but to turn them out to work long before they could either read or write. Eventually by some means or other now forgotten one of them found sanctuary in Calumet, and Hugh Hall's father followed him by the wearisome route across the Isthmus of Panama and then through the gold camps of the Mother Lode in California. Nine years in Calumet accumulated enough savings to take him back to Cornwall in 1867, where he courted Ellen Peters of Redruth and made arrangements for her and her family to join him in Calumet. Their marriage was in the following year and their first house was a log cabin "right out in the woods" where they "had to place logs across the door and windows at night to keep out the bears". Here, enduring the harshness of heavy snows and piercing winds that swept across the wastes, they reared a family in the disciplines of Methodism, the wife teaching the husband to read by studying the Bible. There was never any real shortage of work at the Cliff, the Central and the Copper Falls mines until 1883, when the price of copper started to fall. So the miner moved west in the trails of others: "the railroads were just opening up the farmlands of the West at that time and quite a large colony of laid-off miners from Michigan went to North Dakota and tried farming". But the Badlands proved unkind and their skills were poor. So in 1890, after nine years of wasted effort, the father moved them once again, this time to Jamestown, where he and two of his sons found more regular employment on the railroad. Hugh Wall followed suit and, by dint of studying into the night, became a travelling accountant on the Northern Pacific and then the Great Northern with his headquarters at Portland[2].

1 See Norman Carlisle, 'Challenge of the Solid Silver Island', *True* (July 1962).
2 In 1960, at the age of 92, he visited Cornwall, but the mists were down to the ground and he saw nothing of the Land's End he had longed to see.

His brother Tom's career was typically Cornish-American: first a worker on his father's farm twelve miles to the south of Jamestown; at eighteen one of a construction gang on the Aberdeen, Bismarck and North-Western Railroad; and then a clerk with the Northern Pacific at Mandan and Fargo, all the time attending night schools to improve his education. The American Railway Union strike of 1894 threw him out of work, so he decided to go to college, paying his fees by tending cattle or by working as a janitor for the newspaper *The Fargo Argus*. His choice was wise, for in time he became a reporter for that paper and later for *The Fargo Morning Call* and *The Fargo Forum*, an experience that took him to the very centre of public affairs as a politician. For eight years he represented North Dakota at Washington and for twenty-four years was its Republican Secretary of State; and when he died in 1958 at the age of eighty-nine this son of an illiterate Cousin Jack had served his adopted State longer than any other man. He made a large area of Keweenaw Point into a national park and bird sanctuary. His sense of history led him to publish accounts of the Indians of North Dakota, Sitting Bull and John Grass, and the squaw Sakakawea who was one of the guides of Lewis and Clark. But today he is mainly remembered for his campaigns on behalf of farmers and stock-raisers to control the flooding of the Missouri and to utilize it for electric power schemes. The big Garrison Dam is a lasting memorial to his efforts[1].

The Halls made their escape while the routes were still wide open and before competition from Montana and Arizona was to drain the Upper Peninsula almost dry. The seriousness of the situation may be assessed from the production figures for 1909. In that year Michigan refined only 11,882 lbs of copper for every man employed in the industry, compared with 23,983 lbs for Montana and 26,312 lbs for Arizona. The need to reduce labour costs either by making men redundant or by cutting back wages led to bitter union disputes and the great strike of 1913-4 drove many of the Cornish to Butte, Bisbee, the iron ranges in the east or to Minnesota. Those who remained with the Calumet-Hecla comforted themselves with the fact that wages there were higher than anywhere else in Michigan, and home rentals, food and fuel less costly than in Butte[2]. Nevertheless the drain continued until the Calumet-Hecla, which in its prime had employed 16,000, by 1950 had shrunk to a mere 1,600; while the population of Houghton County had declined from 71,893 in 1920 to 38,711 in 1950.

1 *The Bismarck Tribune*, 5 December 1958.
2 Gates, pp. 128-131.

So now, if you want to visit a mine, you must content yourself with a tourist ride through the workings of the Arcadian where the ghosts of its miners whisper the endeavours of yesterday among the crumbling ruins of the Edwards, the Douglas, the Concord and St. Mary's, the Quincy, the Atlantic, Isle Royale, Huron and Superior. Here in the twisting passages struggled Thomas Stratton from St. Just, one of a family of thirteen who began his heavy pilgrimage on earth as a water-boy after the mine had killed his father, and then became a captain at the Atlantic and the Mohawk, overcoming air-blasts in the one and settling cat-strikes in the other. Here Samuel Polglase from Wheal Vor, near Helston, crawled from the darkness to study at a business college in Indiana and then to manage the silver and lead properties of a Chicago syndicate in Colorado and New Mexico. From the Calumet Charles Retallic blew on his cornet so well that he was enticed to the growing iron town of Marquette to join its band and in time promoted to be the Superintendent of its Light and Power Department. Married to a Richards, whose family was well established on the Keweenaw Peninsula, his daughter was famed throughout Michigan for her fine singing voice. Only slightly less famed was that of the Cornish folk-hero, the basso-profundo Dick Buller, which could penetrate ten levels below ground and carry fifteen miles above ground[1].

James E. Jopling, an American mining engineer of Marquette, has found the Cornish "the most interesting of the various ethnic types in the Upper Peninsula of Michigan"[2], and claims that in 1882 almost all the superintendents, captains and shift-bosses were Cornish, a fact which is confirmed by the 1883 *History of the Upper Peninsula*. Among the leading personalities the Cornish figure prominently. John Hoar, who has already been mentioned as being one of the first to arrive at Eagle River, became interested in transportation. He built a tramway from the Isle Royale mine to the stamp mills on the Portage Lake, organized the L'Anse and Houghton Transportation Company, and bought the steamer *Ivanhoe* for passenger work on the Lake. He was also elected to the State Legislature for the 1873-5 term. William Harris was another "of the pioneer mining men of Lake Superior in the early days of 1846"; to him is credited the mineral exploration of the Canadian side of Lake Superior and the opening of the Bruce mine. He was the discoverer in 1857 of the now famous 500-ton mass of

---

[1] Alfred Nichols, a Cornishman who had been raised at the Central mine and was later Superintendent of Schools at Osceola, told his stories of Buller with such effect in *The Portage Lake Mining Gazette* that his readers believed that he really existed.

[2] See *The Michigan History Magazine* (Vol. XII, 1928), p. 555.

The remains of the working beam of a Cornish pump installed in the Cora Blanca mine at New Almaden, California, in 1873. Abandoned and left underground, it was brought to the surface again in 1964 by Mr. L. E. Bulmore (left of picture) and is now in the Statehouse Museum, San Jose. [*Stelling, San Jose*]

"Crowst-time" (or lunch-time) for Cornish miners at New Almaden in the 1880's. Seated: James Paull, Dick Collins, William Bishop and James Kissell. Standing: Dick Harry, one unknown, J. Andrews and Charles Derby. Note the Cornish dinner-pail and candlestick. [*Bulmore-Winn Collection*]

Pay-day (or dea de raya) for Cornish and Mexican miners at New Almaden in the 1880's. The bearded figure at the bottom left of the picture is the great James Varcoe from Tywardreath and to the right of him is the cashier R. R. Bulmore.

[*Bulmore-Winn Collection*]

Section of the underground workings of the Cliff Mine, Lake Superior, about 1850, showing the horse-whims for raising the copper ore. [Michigan Technological University]

Surface workings of the Cliff Mine about 1850 before they were modernized by the Cornish manager, Joseph Rawlings. [Michigan Technological University]

LINES

On the fearful ACCIDENT which occurred at the CENTRAL MINES, on the LAKE SUPERIOR, on the 22nd of April, 1872; by which ten Miners lost their lives—eight from Cornwall and two from Devonshire.

SAD news from across the ocean we hear,
  Sad news from the Central Mine,
Sad news for wives and children dear,
  Of death in that distant clime.

'Twas ten o'clock on an April night,
  When a change of men took place,
And thirteen miners in the skip—"all right,"
  Down the shaft were lowered apace.

Ten men were on the top of the skip,
  And three seem'd safe within,
When the wire-rope broke with a sudden snip,
  And it fell with an awful din.

Two of the men fell in the level below,
  In the ten-fathom level 'tis said;
Joel Eade had his arms & legs broken like tow,
  Thomas Bone was killed stone dead.

The cries of poor Eade were dreadful to hear
  As up from the shaft they came,
Which fill'd the hearts of the miners with fear,
  As they stood and heard the same.

The guides from the skip fell off with a crash,
  To the fifty-fathom they fell,
And across the shaft they fix'd with a smash,—
  'Twas to several a funeral knell.

The three in the skip alive remain'd,
  Though frighten'd and much alarm'd,—
In that position awhile sustain'd,
  Not a hair of their head was harm'd.

Eight others were kill'd in that dread place,
  Midst broken timber and stones,
With limbs all smash'd and disfigur'd face,
  Torn-off flesh and splinter'd bones.

When the engineer found the rope was broke,
  The whistle he sounded aloud,
To summon some help to the fearful spot,
  Where soon assembled a crowd.

Some quickly descended the fatal shaft,
  A horrible scene to behold;
Eight of their comrades of life bereft,
  Their bodies all bloody and cold.

They might have gone down much deeper still,
  Full eighty fathoms or more;
For the shaft was a hundred and twenty deep,
  Where they sought for the precious ore,

And one by one they were soon brought up,
  And laid in the change-house near,
Till coffins were made in which they were put,
  And sorrow and sighing were there.

To their homes they bore them in sad array,
  For awhile with friends to remain,
And then to the grave on the first of May,
  There proceeded a mournful train.

On the fourth of May, poor Joel Eade,
  Of his terrible wounds did die,
And soon with his comrades side by side,
  In that foreign grave did lie.

The names of the dead we must here record,
  All bred in Old English homes,
With wives and children depending for bread,
  Who now their sad fate bemoan.

From Bovey Tracy, in Devonshire,
  There two of the victims came,—
Philip Roberts who's left a family dear,
  Thomas Champion a young man's name.

Joel Eade from Ludgvan; from Zennor three
  Who first-cousins were well known—
Thomas Berriman & John—two brothers they,
  And also Thomas Bone.

From Callington was Jacob Gray,
  And William Barritt too;
John Ivey from Camborne came they say,
  Edward Thomas from Marketjew.*

The names of those who were sav'd in the skip
  All three from Cornwall came,
And one of them was a Gwinear chap,
  And Edward Treziee by name.

John Pearce from Crowan known full well,
  John Rowe from Camborne town,
And these were spar'd alive to tell
  Of their comrades stricken down.

Four families at Lake Superior live,
  With husbands and fathers gone;
Without some friends their wants relieve,
  How sad in that land alone.

At home there are four more families left,
  No husband or father dear,
By this sad accident of those bereft,
  They lov'd with a love sincere.

Though their graves are made in a foreign land,
  And their forms no more we shall see,
Yet we hope to meet them on Canaan's strand,
  Each one with his family.

For when the last trump sounds thro' the skies,
  Each one shall appear again,
And may they and us with joy arise,
  The Saviour to meet.—Amen.
      * Marazion.

Harris, Printer, Publisher, &c., Hayle.   *Book bound in any style.*

A mining ballad from an unknown hand that, in its rough way, reveals far more than any newspaper account does of the tragedy of emigration. On this occasion all the men killed came from the south-west, from Bovey Tracy in the east to Zennor in the west. Such miners as these are still remembered in the annual pilgrimage to the Central Mine.   [Royal Institution of Cornwall, Truro] copper, said to be the largest ever mined, for it required forty men and twenty months to cut it up. He too interested himself in business enterprises, this time the docks and harbours of Lake Linden; and twice served his time in the Michigan Legislature, from 1871 to 1875. Three others deserve mention among the pioneers. Joseph W. V. Rawlins was a construction engineer at the Cliff and the Bruce a year or so before gold was being panned in California and became chief draughtsman and assistant superintendent at the Portage Foundry and Machine Works. In a different class was Thomas Davey, washing copper at the age of twelve in Cornwall, then at twenty-five for the Quebec Mining Company "in the heart of the great wilderness" until he reached the position of foreman at the Franklin stamp mills.

Those Cornish who were in posts of special responsibility in 1883 had been working over the Peninsula for more than thirty years, unmoved by the lure of the Golden West. William Harris superintended the Centennial. Captain Richard Uren was the secretary of the Lake Superior Native Copper Works with decades of experience behind him as miner, agent and then promoter (with Thomas Dunstone and Joseph Blight) of the Eagle River Fuse Factory. The very first captain at the Central was Samuel Bennetts. Captain Josiah Hall was the inventor of the dumping skip, "now in general use in elevating ore", which he first introduced at the North American mine. John Cliff, captain at the Quincy, was assured "an honoured place among the leading mining men of this area". Richard S. Polglase captained the Atlantic and John M. Richards managed the Ahmuk. William Jacka was the first of the captains at the Copper Falls after similar posts at the Minong, the Isle Royale and the Allouez. James Dunstan was agent at the Central. Thomas Burgan, on the payroll of the Calumet and Hecla, was their foreman in the blacksmith's shop at Lake Linden. John W. Richards, who "commenced with the Calumet on the opening of that splendid mine", was their chief copper washer, while the manager of stores was Edwin Henwood who had crossed the Atlantic when he was one. One emigrant of the 1850's who hid his light under the bushel of Ontonogan was William Lean, for he ultimately excelled as a Judge of Probate and the father of eighteen children.

To some it might appear that these Cornishmen were but little fish in a big pond, yet they maintained the throne of King Copper in Michigan for half a century. William Trebilcock was foreman at the "Hecla tail-house" at Hancock. James C. Trembath did all the paperwork for the Cliff, for he had once been a schoolmaster. John Daniell superintended the Osceola and James P. Richards its coterie of captains, including Richard Bawden at the Phoenix and Thomas Dennis at the Franklin. Emmanuel Richards was in charge of the machine shops at the Franklin, and Edward Trevillion of the copper washing at the Pewabic. A few contracted out of mining altogether: Richard Burge, "who disdains any desire to return to underground work", felt happier as the proprietor of the "leading hotel" in Red Jacket; George Jacka felled timber and sold it to the Calumet; Richard Bastian, two years a sheriff's deputy, managed a boarding house; and William Jewell operated a hotel on Lake Linden. William J. Tonkin, strong in the law after two years as a boss in the iron mines of Peru, put his strength to the exacting test of the post of village marshal and chief of police at Red Jacket. W. C. Kinsman was much in demand as the chief supplier of harness to the mines. Thomas Whittle brought to his captaincy at the Pewabic five strenuous years of mining in South America. But the remainder that the 1883 *History* mentions are now no more than names, thirty or more who, in their day as captains, foremen and superintendents, were unsurpassed for their technical excellence and their administrative ability.

Second only in importance to copper in northern Michigan was iron. On the Marquette Range the first ore, it was said, was found in 1845 under the roots of a fallen pine by a Chippewa near the present Negaunee (later generations of Cornishmen pronounced it Negony to match their own Tregony). Four years later rich deposits near Teal Lake were exploited by the Marquette Iron Company under the brilliant direction of Robert Graveraet and Peter White. They worked against immense odds: sudden storms, lack of roads and supply problems which, together with a shortage of charcoal, cost the company 200 dollars to smelt a ton of iron and deliver it to Pittsburgh where the market price was only 80 dollars[1]. Yet, when Graveraet built his first forge, already "Cornish and Irish miners were piling up tons of ore, black magnetite stuff they called hard ore"[2]. But they were migrants like the English swallow, departing south with the onset of winter and to attract them back to this heavily timbered wilderness of maple and birch, two things were necessary: the building of a plank road from the harbour of Marquette; and the completion in 1853 of the Sault Canal that linked lakes Superior and Michigan[3].

The Cleveland Iron Mining Company made every effort to recruit

[1] Harlan Hatcher, *A Century of Iron and Men* (New York: 1950), p. 45.
[2] Stewart H. Holbrooke, *Iron Brew, A Century of American Ore and Steel* (1939), pp. 29-30.
[3] Hatcher, p. 65.

its miners but by 1859 there were barely 500 in Marquette, Negaunee and Ishpeming, and they were a floating flotsam and jetsam of French and Canadian lumber-jacks except for a handful of German, Irish and Cornish "attracted by the promise of higher pay"[1]. As yet this was no place for the Cousin Jack for the iron was quarried from pits and they despised surface miners as "bloody ditch-diggers"[2]. But as the demand for iron soared with the fortunes of the Union cause, so around the original Cleveland new mines began to cluster: New York, Lake Angeline, Humboldt, Iron Mountain etc. To them came the Cousin Jacks, straight from Cornwall or from the bog-iron and charcoal pits of Ohio and Pennsylvania, to work by the side of the French-Canadians whose hissing saws were cutting wide swathes through the forests to satisfy the furnaces and the mills. By 1870 there were 6,103 miners and lumber-jacks living around Ishpeming, 3,254 in the neighbourhood of Negaunee and 4,617 at the port of Marquette. Nearly two-thirds of the miners were Cornish and Irish, compared with one-tenth who were American and Canadian. Their wage was generally lower than on the Keweenaw Peninsula, only 2.12 dollars a day[3]. But here, as elsewhere, the Cornish were masters of their craft, as one American historian has observed generously:[4]

The Cornishmen were among the greatest of the mining captains, those immensely able underground technicians who were always observing and studying the progress of the workings and seemed, almost by intuition, to follow the crazy convolutions of the rock formations. It was a matter of high personal ambition with them to get out at least expense the maximum of clean ore, to avoid digging through waste rock and to guard the safety of the miners. Experienced geologists learned to listen carefully to the words of these experienced captains on the nature and position of the ore deposits . . . Several of the Cornishmen became famous, even legendary, on the range, for their physical strength, their ability to wrestle, handle a drill or load more ore than the next man.

Such a man was William Stanway from Illogan who in 1957 died at a mining location near Negaunee, to this day known as "Cornishtown"[5]. Another was John Rowe, brought from Cornwall when he was five, Sheriff of Gogebic County, City Marshal of Bessemer and in 1910 the undefeated world champion of Cornish-style wrestling[6]. And one who remembers them all is William H. Richards, himself one of the rolling stones that are forever Cornwall, exchanging the golden sands of Hayle

1 Hatcher, p. 89.
2 Holbrooke, p. 108.
3 Hatcher, p. 117.
4 Hatcher, pp. 142-3.
5 *The Mining Journal*, Marquette, 12 November 1957 and 22 November 1958.
6 Ibid., 2 April 1958.

and the babble of eight brothers and sisters for a twenty-first birthday in the iron soil of Negaunee. He moved on to British Columbia, Montana and Idaho before returning to his second home at Negaunee, where he was elected eight times to the post of County Mine Inspector.

Today at Ishpeming, the industrial capital of Marquette County, the Trebilcocks, the Hendras and the Treweeks are to be found buried alongside the Italian Bertuccis, the Swedish Carlsons and the French Delongchamps. This crowded cemetery, that once received a miner's body every week, drums the beat of the Cornish names: Thomas and Kate Trewartha, Richard and Emily Hendra, Thomas Goldsworthy, Joseph Richards, Stephen and Sarah Collick, Martha Pengilly and Robert Arthur. Only a few remain of that age-group that was born in the 1880's, like Arthur James from Ashton who once drove a horse-bus on the turnpike between Penzance and Helston, never dreaming that one day he would be elected a Justice of the Peace for Wakefield, Michigan, an unheard-of distinction then for a working man in Cornwall. None would claim the mantle of greatness, but their new country offered them new opportunities and they seized them to improve themselves and the quality of frontier life. William Whale, eighty-four years old in 1958, with a hole in his head from an explosion at the Lake Angeline mine, was gratified to see two grandsons qualify as doctors of medicine long before their cousins in Cornwall could do so[1].

A few miles west of Negaunee and Ishpeming is the old mine of Champion, so named by the Cornish who pronounced the quality of its ore to be "champion". In an area that boasted several hematite mines, it was the only one of three magnetite mines (the other two were the East and the Keystone) to survive, according to Wilfred Nevue of River Forest, Illinois, whose grandparents, the Delongchamps, homesteaded from Canada in 1855. Nevue was born in Champion and went to school with Cornish children in the 1880's when the countryside radiated an excitement and a joy that Wordsworth felt in a similar lake country in England. Though in winter perishing winds and biting blizzards scoured the hard ground below the flimsy wooden houses, summer and fall blazed their light on the scores of small lakes and creeks. Here Cornish children fished for speckled trout under dark-green jackpines from banks that were a jungle of alders, tall ferns and vines, and walked barefoot through cool swampland or scrambled among tufts of grass and stretches of sand to a plain, grown over with arbutus, that reached to the very bluffs of the Huron Mountains. It was a land to delight the heart of

---

1   Ibid., 22 May 1958.

any boy, whether Cornish, American or Canadian, who wished to become a northern Huck Finn.

Nevue, former lumberman in the Huron Mountains and on Puget Sound and now a retired attorney, worked for a time at the Champion alongside the Cornish. There were about twenty-five families, he says, who lived close to the mine in clapboard cabins provided by the company. The overlords were Rowes, Pascoes, Simmonses and Slee-mans, all of whom had arrived from Canada by way of Saulte St. Marie and Marquette. They were in a minority among a dominantly French-Canadian community and of course spoke no French. But neither did the "frogs" or "peasoups" speak Cornish, so the two races sat down to tolerate each other; and this was easy for there was a mutual admiration for the skills of the other in removing trees and ore. From the French-Canadians the Cornish learnt the techniques of trapping "snowshoe rabbits" for their interminable pasties, and from the Cornish the lumbermen took many lessons in the art of wrestling. "I was always fascinated", says Nevue, "in seeing those wrestling matches with the participants barefooted and using loose duck jackets; and if the wrestlers were paired off at equal size the Cornish always won". One who, however, disgraced himself, was Jack King, who took part in the great train robbery of 1893 just outside Hancock for which he was sentenced to five years in jail. Today, their reputation at Champion has almost disappeared, along with their barns, outhouses, stores, saloons, the town hall, and their homes and chapel.

Due south from Champion is the equally famed Menominee Range, so encased in iron that almost every township must include the metal in its name, as in Iron Mountain, Iron River and Ironwood. Iron Mountain is perhaps the oldest for in 1954 Lawrence D. Tucker, son of a Cousin Jack and editor of *The Iron Mountain News*, produced a souvenir edition to celebrate the seventy-fifth birthday of the founding of Iron Mountain. Not unexpectedly Cornish names and faces tumble from its pages, as well as an account of an annual softball match between the Swedes and the "Cousinjack" team. Martin Harvey and Thomas Hosking are described as "No. 2 and No. 3 of Iron Mountain's earliest living pioneers". Captain Thomas B. Rundle, born in Cornwall in 1836, is superintendent of the Chapin mine. Pierce Knee-bone is pictured with a bright batch of hounds and hunting guns. Joseph Tippett and Edward Tonkin are blowing their horns in Herman Ohlsen's Cornet Band. John Rule lays the summer's dust with his wooden sprinkling cart and Thomas Uren claims immortality by operating the first "motorised hearse". And in 1879 the gospel was first heard in Iron Mountain when the Cornish miner Richard Cudlip preached in the dining-room of the Chapin boarding house two years before the Methodist Church was built. The Chapin was the most famous of all the mines and Cudlip was its first casualty, killed outright and leaving only one son who was to become the President of the First National Bank in Iron Mountain. From its opening in 1880 until its closure in 1932 men like Cudlip excavated it to a depth of 1,520 ft and extracted over 27,000,000 tons of iron-ore. And here was the mechanical wonder of the day, an enormous Cornish pump that has now been moved from its original housing to be preserved as a historical monument. The most powerful in the United States, it boasted a bob of 120 tons, a flywheel of 160 tons, 40 ft in diameter, revolving on a shaft 27 ins in diameter, and a high pressure cylinder of 50 ins and a low pressure cylinder of 100 ins. Costing 250,000 dollars it could lift 200 tons of water from a depth of 1500 ft every minute or four million gallons a day.

This iron range, like the copper range, has been found a place in literature. Miss Jo Valentine's novel *The Trouble in Thor*[1] takes as its theme the tensions of a mining community near the Wisconsin border. It was "a violent land of no great mountains and feathered by forests", dominated by the straddle of company houses, the wooden grade school and "the little white Methodist Church" round the West Thor mine. The society of miners is feudal, one of services and class servitude, controlled by the Cornish under the leadership of Captain Gideon Trezona. Fifty-five years old and "in his prime", he "bossed through a hierarchy of assistants and miners, drillers, muckers, trammers and timbermen", while at home he moved quietly but authoritatively through his own enclosed establishment of Cornish Methodist friends, refusing to have any social entacts with other nationalities. His son, not inaptly named Wesley, represents, however, the collapse of this autocracy for he falls in love with a Roman Catholic, while the Cornish pay-clerk, Cyril Varker, seems to be the symbol of that darker side of mining life, personal indebtedness. Though to the American superintendent of the mine Trezona must always remain a Cornish mystery, yet his advice can never be ignored, even though it may seem most irrational. For instance, when the order is given to take a second side-slice off a pillar in the shaft, thus weakening the pillar, Trezona simply "sticks his nose in the air and says he doesn't like it". When pressed for an explanation, "he brings in God". It is precisely to justify

---

[1]   Published in 1953 by Coward-McCann, New York.

the ways of God to men that the novel is concerned, for when the management ignore the warnings of the prophetic Gideon, there is a cave-in. We see five trapped miners, one Italian and four Cousin Jacks, including his own son, pinning their faith to Trezona in a black confusion of rock, wood and human bodies, where differences in religious beliefs count for nothing. His first rescue tunnel collapses. The second succeeds but only his son survives, which he regards as the justification of his faith in God; and the novel ends with him reading from the Bible, "Rejoice with me for h'I 'ave found my Sheep which was lost". With his faith in the God of the miner vindicated, the Cornish way of life stands revealed as stable, sensible and logical, even though its mysterious power cannot always be understood.

No account of the Cornish in Michigan would be complete without some mention of their pastors, missioners and social workers, for in no other mining state were they so actively engaged on their errands of personal salvation. H. J. Nicholls of Beloit, Wisconsin, son of an Ironwood Cousin Jack from Penzance, was for thirty-seven years a field officer of the Salvation Army. Percy A. Angove, a copper miner from Hayle, became the Executive Director of the Michigan Society for Crippled Children and Adults, after a successful college course at Western Michigan University at Kalamazoo. But most astonishing has been the career of Frank Dyer, Doctor of Divinity, born in 1875 in the legendary kingdom of Mark of Cornwall where it encompasses the village of Golant high above the Fowey, loveliest of Cornwall's rivers. "Being the eleventh child", he says, "it is hard to imagine any great welcome awaiting my arrival", but he was not the last for two more arrived, another garden had to be spaded to cope with the fertility rate, and those children old enough were turned out to earn their pennies. Young Dyer was lucky in as far as he was placed "in service" with the Rashleighs of Menabilly, one of the aristocratic families of Cornwall, who in 1893, when he was seventeen, helped him with his fare to Chicago. Already quite clear in his mind about his life's work, he enrolled at the Moody Bible Institute and then qualified as a Congregational minister. His work was broadly evangelical for it was based on childhood experiences at Golant, where his family divided their devotions between St. Sampson's Church and the Wesleyan Chapel "where lay preachers preached the Gospel in plain language". It began in Iron Mountain and finished in Los Angeles, where he founded the fine Congregational Church on Wilshire Boulevard. In 1947 representatives of many faiths came there to present their pentecostal flames of praise after his fifty years of service in the brotherhood of men. His achievements were impressive: pastor of seven churches; guest preacher to five theological seminaries and eight colleges and universities; President of the Pacific South-West Theological Conference, of Junipero's Preaching Friars, and of the Ecumenical Council of Churches; and finally Ecumenical Bishop at the outbreak of the second World War. But perhaps the call he liked best came in 1926 when he was invited to preach in London for this gave him the long-awaited chance to visit Golant, where the finest thrill of a lifetime of preaching was to stand in the pulpit of the church of St. Sampson.

Copper-rich stopes and wooden pulpits in Michigan always resounded to the heavy tread of Cornish boots for they were inseparable from the full life they tried to live; and to nowhere else in North America did Cornwall export so many pastors and preachers. The minutes of the Detroit Conference reveal the obituaries of scores of ministers who began life as miners and then went to college: Lewis Keast, Joseph Oatey, Richard Wyatt, Sidney Eva, John Bunny, William Clemens Clemo, Stephen Polkinghorne, Richard Carlyon, Angwin Rowe, James Ivey, Justus A. Rowe, Thomas Cox, Henry Rogers, Timothy Edwards, James Pascoe, David Shugg, Joseph Pengelley, Joseph Chapman, Paul Nicholas, Samuel Tamblyn, Edward Hocking, Joseph James, William Phillips, John Dingle, Edwin Stephens, James Chapman, John Williams, Waldren Geach, Thomas Bottrell, Lawrence Worth, William Richards and George Prout. The tide of rural Methodism in Cornwall was strong and powerful, carrying away many a miner to train in American universities and colleges because of the lack of opportunities in England. One was James Hitchens, in 1958 pastor of the First Methodist Church at Ottawa, Illinois. He was only nine years old when his mother took him to the Transvaal to join his father, the manager of the Rose Deep gold mine. He mined at Trimountain and then, on his savings, studied at Northwestern University. But their leader until he died in 1962 at the age of ninety was John Strike from Porthleven who, by his enthusiasm and persistence, encouraged many others to follow him in the Ministry of Christ: Herbert Hichens of Newlyn, Harry Colenso of Helston, William Combellack, Romily Prouse, William S. Pryor, Athanasius Rickard, Samuel Bottrell, Howard Snell, Ethan Bray of the Lizard (both of whose sons had pastorates in Flint) and Dr. William C. S. Pellowe, minister of Adrian Methodist Church and College, District Superintendent of Saginaw, an emigrant from the narrow-streeted Penryn, a brilliant preacher and administrator, humorist and raconteur[1].

1 See his *Chuckles in the Cemetery*, *An Anthology of Michigan Poets*, etc.

John Strike's family had little affinity with mines, though their occupations were not less dangerous. His father was a fisherman who nightly set out with nets and lines from Porthleven's triple harbour of granite that is pounded by the English Channel. They were so poor that as a child John Strike worked for two sawyers, driving wedges, and only attended school when they could not employ him. When he was fourteen his father died, so for the next seven years he helped to maintain his mother and three other children by going to sea in the pilchard boats, at first earning no more than three shillings a week. But the nightly tossing on the water was an experience that changed his whole life for, as he remarks: "I have thanked God many times that in my young manhood days I was led into a group of Porthleven Christian fishermen. I was converted in the Fore Street Methodist manse at the age of sixteen and started out in Christian work with the Mission Band".

From the moment of illumination when he knew that he was to be a fisher of men, a wide world of redemption was revealed. In the small library of the chapel he read stories "about the children in San Domingo and Hayti"; and in the Mission for Seamen off the dark streets of the Barbican at Plymouth, where the Pilgrim Fathers had once embarked for an unknown America, he browsed through a "world's literature", provided by that secular saint of sailors and seamen, Aggie Weston. Fisherman for the last time on Good Friday 1893, Strike landed in Michigan that summer to join the copper miners and put himself on the road to ordination. For two years or more he hammered cheerfully in the Ishpeming mines until he had saved enough to begin a course of study at Albion College; this was to last a full seven years and had to be paid for by vacation employment like peddling stereoptician views from town to town on his bicycle, his wage being forty per cent of his sales. Many were the Cornishmen who passed through the lecture rooms of Albion on their way to the mission-field; three of Strike's contemporaries were Thomas Martin, one of the first Methodists in the Philippines, James Trewhella who qualified as a medical missionary and William Connibear, a prominent mining engineer on the Marquette Range. Strike graduated in 1905 and was despatched to the edge of the Huron Mountains among the forested mines of Champion and Michigamme. It was a home from home, for two of his helpers were Herbert Hichens, a pump man from the fishing village of Newlyn, and Captain William Williams from the golden shores of Carbis Bay. Apart from these seven years of plenty on the iron range, Strike's main work for almost half a century was where the Cornish settlement of Michigan first began, on the Keweenaw Peninsula at Calumet, where his Centennial Church served a population of almost 65,000.

Another "athlete of Christ" was William Pearce. Although his ministry does not strictly belong to Michigan, for his province was the whole of America and the Pacific beyond, yet it was on the Upper Peninsula that he started the studies that enabled him to become a Bishop of the Free Methodist Church[1]. He is buried in New York's Riverside cemetery but was born at the height of Civil War at Townshend, one of the many little mining settlements in west Cornwall; he was the youngest of the ten children of John Richard Pearce and Ann Bowden Hoskins. Of his early life we know little except his memories of the Methodist revivalist campaign of 1882, as a result of which he was "sanctified wholly nine months afterwards in the brushwood skirting one of my father's fields through reading 'The King's Highway' ". Probably Sundays saw him as a local preacher in the scattered chapels where the small congregations shrank smaller as the young men marched off to America. So in 1884 he too joined them in Michigan, working the iron-heavy soil for almost a year and at the same time wearily reading for his ministry. Two years later he was ordained in California and began his preaching on foot and from wagons on the Mother Lode with such success that he was promoted Elder of the Ione and San José Districts. In 1900 he was transferred to Oregon to administer the Portland and Salem Districts and then in 1904 was appointed pastor to the Church of Jamestown, New York. His pioneering days in the West were over, but they had proved him to be a capable leader, an efficient administrator, a powerful preacher and a successful evangelist. Out of the heat and the dust of the Sierras and the desert had come the rock on which to build anew the church in the East. Within a year he was District Elder of the Genese Conference and three years later its Bishop, and he was still only forty-six; and for thirty-nine years he remained Bishop, a record in the history of Free Methodism. In 1910 he represented the American Methodist Church at the World Ecumenical Missionary Conference at Edinburgh and seized his chance to visit Townshend and preach in its unpretentious chapel after an absence of twenty-five years, finding to his delight that he had not forgotten his Cornish dialect and resolving that in his retirement he would compile a Cornish dictionary. After the first World War he was attending a similar conference in Japan, from where he intended to

---

1 See *The Free Methodist*, 3 October 1947; and Richard Blews, *Master Workmen* (Centennial Edition: 1960).

visit the missions in China, but was deterred by the revolution there and the mass exodus of field workers from Honan.

His own erudition, which he "wore as lightly as a flower"[1], was known only to those who sat in the pew or gathered in the revivalist tent, for he published only a handful of articles and a course of sermons: *Our Incarnate Lord*. But once he had ascended the pulpit, out rolled the gripping phrase that could elevate his listeners to the heights of the sublime, no matter what their background, for "his utterances had a literary finish which charmed the educated, coupled with a simplicity and a conciseness which held the uneducated". A phrase from Tennyson, a tale from Longfellow, a moral from Bunyan or a line from Milton would illuminate the flow of ideas that came from years of disciplined study, but illustrations would also be drawn from the language and the situations of ordinary men and women, for he too had laboured with his hands. Thus his "mighty intellect which roamed the universe" was graced with a humility that stemmed from tiring muscles and an aching back. Of a simple piety and an exacting faith, combining the austerity of the Puritan with the aestheticism of the romantic, he was remarkably gifted for his handling of men. In the swamps of argument at conferences his flash of humour and his quick repartee would disperse any clouds of misunderstanding and make an adventure of the pedestrian routine of business. The source of his moral strength, he said, was "sanctification by faith" and "a sense of sin (that) bore down upon me with violence as the terrific pressures of the intensely spiritual revival atmosphere supervened". This moral strength enabled him to rise above the two tragedies of his personal life. His first wife was killed in a riding accident after nineteen years of married life just three months before he was consecrated Bishop. His second wife died in giving birth to a daughter. Now a well-known organist in Philadelphia, she likes to visit Townshend where her father lived in the eight-roomed granite house that still stands occupied. It is still called Kirthen Water and it looks out upon the same brushwood where, all alone, he saw a vision of his ministry in America. "Celtic by blood, Cornishman in particular, Britisher by birth, American by adoption, Christian by second birth and saint by processes of grace and experience", as Bishop Fairburn once said of him, William Pearce must have seemed a veritable candle of the Lord to those who knew him. Today and tomorrow he will be remembered on the campus

---

[1] See William Pearce, 'The Preacher and his Reading', *The New Methodist* (31 May and 7 June 1935).

of the Roberts Wesleyan College at North Chili in New York State where, within a new church of red brick fronted by white columns and surmounted by a slim white tower in the colonial style, is the Pearce Memorial Hall.

Some Cornish pastors in Michigan followed their flocks along the southbound trail that led to the forbidding Dakotas. Here, as Hamlin Garland wrote in his *Main-Travelled Roads* of 1887, "the houses, bare as boxes, dropped on the treeless plain, the barbed-wire fences running at right-angles, and the towns mere assemblages of flimsy wooden sheds with painted pine battlements, produced on me the effect of an almost helpless and sterile poverty". For the miner, however, the Bad-lands terminated in the Black Hills where, ten years earlier, the Home-stake Ledge of gold had been found. Today the Homestake Mining Company remains the major producer of gold in the United States.

It is a company that can well boast a proud record as an employer of labour for the Cousin Jacks who went there to slake their thirst for ore usually stayed, so bright were the prospects and the conditions of service[1]. The two main towns of Deadwood and Lead, surrounded by deep forests abounding in deer, elk and antelope, were always dominantly Cornish. At the time of the 1880 census seventy-five of them with their families had settled in, ready to welcome others from Michigan or Cornwall. James Blamey Harris came in from Truro in 1885, married a Reynolds and stayed for the remainder of his life; Matthew Furze from St. Just-in-Penwith settled in Lead in 1905 with a Cornish wife who had appeared in many a stage production at the Cornish-built Opera House in Central City; Isaac Nichols without regrets forsook the stony slopes of quoits and iron-age forts around New Mill and Madron, mined at Calumet, returned to Boskenwyn for a wife and then married her in Lead; and Phillip Cann and his wife from Chacewater lived within sound of the Homestake stamps for forty-two years, and their son after them.

"It is a good outfit to work for", writes S. J. White of Lead, who remembers the irritant poverty of his childhood when his father, a hoist engineer at the Prince of Wales mine at Callington, was widowed and on his own struggled to provide for his ten children. "How did I become an American?", he says, "That is not hard to explain. I think the north east section of Cornwall was one of the poorest spots in the whole county, for jobs were hard to find, the small mines only operated in spells, and the farms were of the 'wheelbarrow' type". He adds,

---

[1] See *The Story of Homestake* (The Homestake Mining Company: 1954).

significantly: "I don't think there is a country in the world where a man has to work harder for his wages than in the U.S.A.".". But in the clay pits of St. Austell a man would have to work very hard to earn as little as twenty cents a day, so clayworker Arthur Morcam dragged himself off to Butte, arrived in the middle of a depression, moved on to Houghton, was given a job by the Cornish captain at the Quincy (a Kendall from Roche) and then, with Philip Dyer (whom he had known at Penwithick), came to rest in the Homestake, no more to shovel muck but to direct men on the shift.

The Dakotas were spiritually serviced by three remarkable men from St. Day. The first was John Hall, the son of a lay preaching mine captain, who in 1882 was among the waste-gang at the bottom of the Homestake, earning 90 dollars a month compared with the pittance of 10 dollars in Cornwall. While working a night shift he felt a call to train for the Ministry, studied for six years at Black Hills College in Hot Springs and then at the Garrett Bible Institute, and paid his fees and maintained a wife by odd vacation jobs like sheep-herding on the Cheyenne River. For ten years they rode the dust of the Black Hills, raising money at Lead and Deadwood for the repair of delapidated churches and infusing new enthusiasm in congregations contented to make do with services in a Fireman's Hall[1]. The work was exhausting, unremunerative and disappointing. There was no manse; and Methodism was grinding to a halt. Needing more power to his elbow, he wrote to a promising young preacher in the parish of Gwennap, Will Richards, who was working seventy-two hours a week in an ironmonger's shop for a paltry wage of two shillings a week. Hall assigned him to the Cousin Jacks at Terraville on the site of the Homestake's stamp mills, where there was no observance of Sunday, and every day was noisy with the exploits of "Calamity Jane", the mail carrier between Deadwood and Custer, and of "Deadwood Dick", the Cornishman Richard Bullock who rode shotgun for the Homestake[2]. Circuit rider between Spearfish and Belle Fourche, Richards despaired of cowboys who never went to church, but was encouraged by the advice of the Bishop who ordained him: "A Methodist preacher must remember two things—he must be humble and he must be poor. The Lord will attend to his being humble and the congregation will attend to his being poor".

---

1 John Hall died in 1960 on the eve of his 90th birthday after 40 years service in the Minnesota Conference.
2 See *The Old Cornwall Magazine* for the summer of 1943. There have been at least three claimants to the title, the latest being that of one Richard Clark who was buried in Deadwood as recently as 1950.

In 1960 at the age of eighty-four he was still not too old to serve the church at Skamakawa on the Columbia River and to visit Cornwall.

The third of these three evangelists from St. Day was Tom Burden, orphaned in Redruth when he was a year old and brought up by a relative in nearby Carharrack. He emigrated to North Dakota on the invitation of a Cousin Jack who had heard him preach, but the way ahead to the ministry was hard though not unusual. He trained at the University of North Dakota, Wesley College and the Boston University School of Theology, keeping himself peddling pianos, a vacation job which must have required no small gifts of persuasion. Thus doctrinally and spiritually equipped, he won his spurs in California: at Fresno, Tulane, Inglewood, Annaheim and finally Los Angeles and Burbank, where the library of the First Methodist Church bears his name.

V

COLORADO AND "THE RICHEST SQUARE MILE ON EARTH"

> Uneven crags and cliffs of various form;
> Abysmal depths and dire profundities;
> Chasms so deep and awful that the eye
> Of soaring eagle dare not gaze below
> Lest, dizzied, he should lose his aerial poise,
> And headlong falling reach the gulf below.

So Alfred Castner King remembered the mountains of the San Juan district of Colorado as he lay in the darkened room of a Denver hospital. Not yet thirty and the son of Cornish immigrants from Michigan, he had come to Colorado to hunt for silver, like so many of the Cousin Jacks on Lake Superior, and had established himself as an assayer at Ouray. In March 1900 a box of dynamite caps exploded in his face with fearful consequences; a beard had to be grown to cover his disfigurement and he was totally blind. Yet from this tragedy emerged a man who tried to recapture in words the sensations he had once experienced of the gold and silver canyons as he viewed them from the top of Mount Wilson.

His two volumes of poems, *Mountain Idylls* and *The Passing of the Storm*, are dedicated "to a rapidly disappearing class, the pioneer prospectors, whose bravery, intelligence and industry blazed the trails in the western wilderness for advancing civilisation and made possible the development of the Great West":

> Some blest of wealth, some cursed by poverty;
> Some in positions neutral to them both;
> Some wore a gaunt and ill-conditioned look
> Which told its tale of lack of nourishment.

Others too merit his compassion: the man "of large affairs", the "brainless fop", the "man of letters", the "gay coquette"; and above all the miner who

> Ere the shift is done may be crushed or broke,
> Or the Life may succumb to the gas and smoke.

Yet all those engaged in one of the greatest mining scrambles he reminds of a warning from Oliver Goldsmith: "Ill fares the land, to

hastening ills a prey, when wealth accumulates and men decay". So, in
The Nation's Peril, he reveals himself as a Christian socialist, concerned for a just society based on a reasonable distribution of opportunities:

> The heritage of man, the earth,
> Was framed for homes, not vast estates;
> A lowering scale of human worth
> Each generation demonstrates,
> Which feels the landlord's iron hand,
> And, hopeless, plod with effort brave;
> Who love no home can love no land;
> These own no home until the grave.

A natural poet who sang his way through the drab world of the miners
with ballads dressed in a Miltonic solemnity, King possessed no more
literary background than that provided by grade schools and the
libraries of the miners' institutes. Yet from his gift of spinning words
and weaving rhythms he made a living, lecturing in America and
abroad on his experiences in the mining camps. Harold Axford, associate
editor of *The Oregon Journal* and grandson of a Cousin Jack from
Redruth who worked in the coalfields of Pennsylvania and then on
the Denver and Rio Grande railroad, remembers "Cassie" King well
for he went to school with his father. When he was peddling his poems
from town to town he would often call at the Axfords' home, where
he would delight the family with tunes on his flute and his Cousin Jack
stories, some of them his own invention; but he would never permit
them to be written down[1].

Through sheer size, the Rocky Mountains that stimulated King's
poetic imagination are less easy to comprehend than the Sierra Nevada,
and they are more varied and spectacular, though less dramatic; for
whereas the Sierras are a solid wall rising steeply from the desert
and prairie, the Rockies form a broad series of ranges. The one was
a barrier while the other is a drainage divide, for in the Rockies wind
and water have severed the masses to form passes and gulches. Thus
the immigrants to the Pacific West were provided with natural passage
ways that linked prairie to desert, in spite of the sheer hulk of the mass
above that reached to the sky, and hurried on. It was not until 1859 that
prospectors from the east, from Virginia and from Georgia, pitched
their tents in what was known as "Kansas Territory in the Rockies"
and stayed. Even today as the Californian Zephyr snakes its way
through hundreds of miles of canyons, their fantastic cliffs and pillars
carved into many-hued strata of ancient sediments by the Colorado

[1] Castner King died in 1941 and is buried at Fruita, Colorado.

River, one is hardly aware of being in the heartland of the Rockies until the train has crawled to an altitude of almost 10,000 ft, penetrated the six miles of the Moffat Tunnel and begun its spectacular long glide down to Denver. Not until the crest has been cleared and the eye has taken in the immensity of the plain below is it possible to understand the thunder and lightning of King's images. Denver rides a mile high in the sky but, because it nestles against the mountain foothills, it appears to be at sea-level. Indeed, when viewed from Boulder on a rainy day as the clouds roll down Pike's Peak, the prairie takes on the appearance of a bay of the ocean and the skyline of Denver becomes the masts of ships at anchor. Denver is a graceful city of lakes and parks that has an English look about it due, not to the British Consulate, but to its houses of warm red brick.

One of them, 1720 Sherman Street, had once been considered one of Denver's historic mansions, for it was the family home of Richard B. Pearce who has been called the "saviour of western mining", and not without reason[1]. Born at Barippa, near Camborne, in 1837, he studied mining at the Royal Institution of Cornwall in Truro and then at the Royal School of Mines in London. From 1865 to 1871 he was manager of the silver and copper smelting works of Williams, Foster and Company at Swansea, and then he was sent to Colorado by a group of English capitalists to modernize their silver mines at Georgetown. But he stayed there, sent to Cornwall for his wife and family, and became superintendent of the Boston and Colorado smelter in Denver. There, in partnership with N. P. Hill of Black Hawk, he introduced new methods to separate gold from its associate metals that soon made Central City famous as the Little Kingdom of Gilpin, proof, it has been said, "that destiny had ordained his many years of arduous training in his native country for great and certain results in her appointed time and place"[2]. He then went on to discover uranium and radium on the dump of the Wood mine[3]. He published many articles, including an account of the state of mining in Cornwall. He was a founder and President of the Colorado Scientific Society, a Fellow of the Geological Society of London and President of the American Institute of Mining Engineers. Columbia College conferred on him the honorary degree of Doctor of Philosophy, eulogising him as "one of the first, if not the first, gold and silver metallurgists of the

---

1   E. E. Kohl, *Denver's Historic Mansions* (Denver: 1957), pp. 27-32.
2   William N. Byers, *The Encyclopedia of Biography of Colorado* (Chicago: 1901), Vol. 1, pp. 233-5.
3   *The Mining World*, September 1959.

present time". That he possessed other sterling qualities seems certain from the action of successive British Governments who appointed him their consul for almost twenty years, an honour without parallel.

The Cornish in the Rockies have been the subject of several studies by American historians. Lynn Irwin Perrigo in *A Social History of Central City* (an unpublished Ph.D. thesis for the University of Colorado, 1936) and in an article, *The Cornish Miners of Early Gilpin County*[1], draws attention to their distinctive social qualities that gave some social stability to the mining camps. Part of their attractiveness he found to be their unusual conventions of speech, transposing pronouns and verbs as well as dropping their aitches in expressions like "'ow art e getting on you"; and they would never "go out" so much as "go forth". Dress seemed to match the language, the women in flamboyant gowns of purple velvet and large hats with curving yellow plumes, and their men in loose-fitting coarse suits and bowler hats. In spite of their verbal illiteracy, however, they were always to be found "mingled in the select social set of Central City", no doubt due to their exceptional singing voices which dispersed all social handicaps and educational short-comings. David H. Stratton in *The Cousin Jacks of Caribou* (published in *The Colorado Quarterly*, Spring 1953) lays emphasis on the encompassing influence of the Methodist Church, where all took a turn as janitor, knocked on street doors to solicit donations towards the minister's stipend, and on Sundays indulged in "fervent prayers" and "friendly handshakes". Miss Caroline Bancroft in *Historic Central City* and *Gulch of Gold*[2] reckons that the Cousin Jacks brought "a sparkle and fun entirely lacking before" and that "they are generous, humorous, spirited, imaginative, superstitious and clannish", but "extremely reserved and not easy to know". Thomas Mitchell of Russel Gulch once told her that there were two kinds of Cornishmen: those who left Cornwall, the Cousin Jacks; and those who stayed behind, the Cornish.

One Cornishman who became a Cousin Jack was the blacksmith from St. Just, Thomas Henry Blackwell, whose story is told by his American-born daughter of Denver, Lilian Blackwell. Economic despair drove him out for he had seen both his parents and a brother and sister waste away and die of consumption. The Methodist Church alone seems to have kept him sane for there he was "leader of our principal Sunday morning class, one of the chapel stewards, one of the principal men in

[1] *The Colorado Magazine*, May 1937.
[2] Published in Denver 1957 and 1958.

the School, and the old stand-by in the choir"[1]. One sister he had helped to emigrate to Australia, and then he made the hard decision to go himself, hard because he had been married only five months and his wife was pregnant. So in the summer of 1876, with his wife's brother and father as companions, he apprehensively headed for California, finding work at Eureka Mills and living through two bitter Sierra winters when the unfamiliar snow was so deep that their only exit was through the roof of their cabin. Here his wife and the daughter he had not yet seen joined him and here another daughter was born. But now they were all making so successful a living that his wife was able to go back to St. Just with her children to bring out the remainder of the family. It was while she was there that the news reached California of gold strikes in Colorado, so the men were soon on the move again, this time to the dizzy heights of Leadville, where Mrs. Blackwell joined them.

Here they battled for five years against the lassitude of an altitude of 12,000 ft until the severity of the winters forced them to search for work elsewhere in the mountains. Mrs. Blackwell's brother was the first to go, leading his family in a wagon over the treacherous 12,200 ft Independence Pass down to the more genial slopes and streams of Cattle Creek, near Glenwood Springs. They built a log ranch house and hewed and shaped their furniture from the living trees; and in due time by their combined efforts more land was acquired until the three sons possessed ranches of their own. Thomas Blackwell followed them in the fall of 1886 and at Aspen rented for 25 dollars a month a two-room log cabin where twin boys were born, only to die before they were a year old, victims of pneumonia that ravaged the mountain camps in the winter. Here he was joined by cousins from St. Just, James Thomas and his wife, but disaster struck them down almost as soon as they arrived, the husband being killed in the mine, and his distracted wife returned to Cornwall. Almost at the same time his father-in-law died, the companion of all his wanderings, mourned by his widow in St. Just who had resolutely refused to join him in the land that lay so far west of Cape Cornwall.

Aspen, a veritable queen of silver, rested on a wide plateau at the base of Red Mountain whose peak reached skywards to a height of 10,000 ft and over which several Cousin Jacks from Penzance had made their perilous passage from Leadville on foot, on burros, in ore wagons and sometimes in Kit Carson's six-horse stage coach. But great as was its fame, it had to yield pride of place to Central City, "the richest

---

[1] Letter from Rev. T. E. Mundy of St. Just, 4 June 1876, in the possession of Miss Blackwell.

square mile on earth". The present highway to it from Denver follows the curves of Clear Creek so closely that the traveller would never guess that this was not the original way the miners came. Until 1878, when the railroad was built, the way into Central City was from Idaho Springs along a perilous shelf-road of dust or mud that twisted and turned for nine miles up Virginia Canyon, and is still a scaring natural roller-coaster. Idaho Springs was a sort of base camp where the miners rested before the grand assault at boarding houses maintained by Cousin Jennies like Selina Bickford and Edith Brent from near Calstock. There is a cemetery nearby with an inscription, as startlingly remote as the disaster it commemorates, in memory of Shadrach and Harry Gale from the same village as Selina Bickford: "Perished with the *Titanic*, April 14th, 1912". When Central City began to die the railroad died too, but in the depression of the 1930's there came an exciting revitalization when the idea was conceived of making the old mining town the cultural capital of Colorado. But tourists were just as reluctant as the old teamsters to use the sky-high Virginia Canyon route because of its dangers. So the tracks were torn up, the modern highway climbed into the railroad and Central City was saved.

From Golden, where today the Trezises farm the land of the Lawrys from Nancherrow, this highway ascends Clear Creek Canyon and leads to the fabulous Russel Gulch, named after the man from Georgia who with his Cherokee Indian wife made his strike in 1859, and then to Black Hawk, first known as Gregory Diggings after John H. Gregory, also from Georgia. As the visitor from Cornwall explores these and other gulches, he knows at once that he is walking through a Cornish town for nowhere else, except in Wisconsin, is the hand of the Cousin Jack so patently visible on the man-made landscape. It is not in the cemeteries of the Knights of Pythias or of the Rocky Mountain Lodge No. 4 of the I.O.O.F. It is not in the Methodist Church of St. James with its chapel dedicated to the Lawrys and the Richardses. Nor is it in the Miners' and Mechanics' Institute of 1871 with its well-stocked library and reading room. What makes the gulches so Cornish in appearance is the dry-stone walling.

"How the Cornish could build dry stone walls, using no mortar, that would stand the onslaught of Rocky Mountain weather for nearly a century, no one knows", comments one American historian[1]. The answer lies in Cornwall where, over the centuries, the craft of wall building had developed to protect a traveller or crops against the

[1]   Caroline Bancroft, *Historic Central City* (Denver: 1957), p. 35.

battering of the Atlantic gales. The masons were the miners, and not unnaturally so, for the two crafts are related; the nature of the granite has to be understood, whether for the purpose of blasting it to extract a mineral or to shape it into a wall. In Cornwall today there remain very few craftsmen who can build these walls, so making it a most expensive form of hedging, but in Central City every miner was an expert mason. So, equally at home with pick and trowel, they terraced the roads and pathways to their houses on the steep hillsides with strong retaining walls and, moreover, when off duty at the mines, built the Teller Opera House, the Episcopalian Church of St. Paul's and the Methodist Church. In a way these are the most enduring signatures of the Cornish miner for there is a timeless quality about them, as unchanging as the men who erected them, as one observer has noted:[1]

The miners, who are perhaps the most numerous class, represent all nations, but after a residence of a year or two in the mountains, lose old national characteristics. The Cornishman changes but little . . . .

Another building in Central City rich in its Cornish associations and sited directly opposite the Opera House is a livery stable which provided, not only feed for horses, but "carriages to Idaho Springs, Pine Creek and Yankee Hill", and was the terminus for the local "City and Transfer Hackline". Today it is used for dance exhibitions during the summer festival of opera, but it was first built to provide a service for the patrons of the Teller House. Then in 1880 it passed into the possession of the "famed Gilpin County Sheriff Dick Williams", one of Central's most respected citizens who met his death one April morning before the revolvers of a gunslinger.

Cornwall perhaps was no more than a name to Williams for he was only a child when his parents arrived in the forlorn regions of Eagle River. By the time he was seventeen he had prospected along the gulches in Gilpin County and saved enough to fetch them. At twenty-six he was elected Mayor of Nevadaville; and at thirty-one Sheriff of Gilpin County, a post he held for six years when "his name was a terror to evil-doers". It was while he was sheriff that he built up his livery business but, after his term was over, he moved to Denver, where he became an equally successful stockman. Three years later in 1891 he returned to Central, was elected County Commissioner for three years and then, in April 1895, Mayor of Central. He had just relin-

---

[1] S. S. Willihan, *The Rocky Mountain Directory and Colorado Gazetteer* (1871), p. 112.

quished this post at the age of forty-eight when the shooting affray occurred that cost him his life.

Samuel Covington earned his living in Nevadaville hauling quartz, but little was known about him except that he wore a belt with pockets under the arms for the concealment of guns and so arranged that they could be drawn in a flash. On the morning of 15 April 1896, incensed that Judge Sebright had ordered the settlement of bills to be paid out of his wages, he burst into the Courthouse, pulled a gun, pressed it against the Judge's chest and fired; fortunately the Judge pushed the gun aside a fraction of a second before it was fired and the bullet lodged in the floor. Marshal Keheler however was shot dead and then Dick Williams appeared, having borrowed a pistol from Elias Goldman's saloon. He too was gunned down, a massive .48 bullet ploughing its way through his ribs and lodging in his stomach with shattering effect. Covington was finally despatched by the Winchester of a cattle-raiser at the Corydon mine, but Williams died four days later, conscious to the end. Never before had Central City witnessed so impressive a funeral, for no church was large enough to contain all the mourners, all shops and mines being closed to allow the workers to attend. So the deep, wide stage of the Opera House, which had opened with *The Bohemian Girl* and where Edwin Booth had acted, became for once a high catafalque for the body of the Cornish hero. Even so, many had to remain outside, and it took almost an hour for them to file past. The procession was the longest ever seen in Gilpin County, 116 carriages, twenty-seven horsemen and an uncounted number on foot of the Sons of St. George from Denver, Boulder, Idaho Springs, Georgetown and Golden. And among the pall bearers were three Cousin Jacks: John Pengelly, William Mitchell and John Teague[1].

When Williams came to Gilpin County in 1865, the fever of the initial rush to Pike's Peak had long since spent itself, and the inexperienced diggers returned to the east. Cousin Jacks are not likely to have been in this rabble that *The Nebraska City News* of 14 April 1860 reviled as a "scurvy horde of pauper emigrants", animated by "reckless folly" and "crazy madness" who, without a dollar in their pockets and provisions to last them only a week, tried to journey for almost a thousand miles over a wild and inhospitable region. Not for him the life of tenting and cooking in the open among the mass of humanity that crowded into Gregory Gulch in the summer of 1859 and succumbed to the "mysterious mountain fever" brought on by the filth that accumulated round the

---

[1] *The Central City Register-Call*, 17, 24 April 1896.

wagons and polluted the streams[1]. Not yet for him the decision whether gulch government was to be that of the knife and pistol or of properly constituted assemblies after the manner of miners' courts, complete with

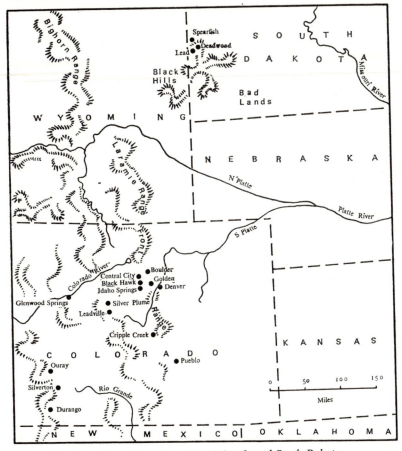

Main Cornish settlements in Colorado and South Dakota

presidents, judges and sheriffs. Although Central City was already being laid out halfway between the two camps of Nevada and Black Hawk on the suggestion of William N. Byers, the founder of *The Rocky Mountain News*, there were too many gigantic failures due to "incompetent mining captains and mill men, swindling mine-operators, buncombe companies with penniless directors and senseless agents, charlatan metallurgical professors with their worthless processes"[2].

1 Caroline Bancroft, *Gulch of Gold* (Denver: 1958), p. 55.   2 Wallihan, p. 46.

So the mine owners deliberately started their quest for Cornish miners, wherever they could be found. Many were the inducements dangled before the Cornish, and not least in order of importance that Gilpin County boasted so mild and healthy a climate that "no unpleasant effluvia is detected in the neighbourhood of dying carcasses" and that "sloughing or indolent ulcer rarely follows gunshot wounds"[1].

More now was required in the gulches than when the first prospectors found the gold coarse and free in the quartz. By 1863 the weathered surface quartz had become exhausted, shafts had to penetrate to deeper levels, and steam power was needed to drain the workings just at the time when the timber was showing signs of exhaustion. Cordwood had to be hauled from Fall River, twenty miles away; the ores were proving to be uniformly refractory; and the stamp mills could do no more than save about 25% of the gold, while at the same time wasting the entire silver and copper content of the ore. What Gilpin County clearly needed desperately was a working party of sound business men, experienced metallurgists and skilled mine captains. If there were any doubts harboured about the Cornish they were speedily dispersed in July 1863 by Maurice O'Connor Morris, at one time the deputy Postmaster-General of Jamaica, who spent several months in Central City collecting material for a book published in the following year, *Rambles in the Rocky Mountains*. He told the mine owners that an influx of Cousin Jacks could transform the whole economy of the gulches[2].

By 1870 this indeed had happened. N. P. Hill was developing new processes at Black Hawk for the recovery of the refractory ores, making the long journey to Swansea to see how he could improve the smelting and returning with a representative of the Cornish firm of Vivian who would arrange to send "matts" of the unseparated metals back to Wales. Cousin Jacks were springing up in the canyons, organizing their choirs in the Montana Theatre, talking about stulls, sprags and cabooners (their own variation of bonanza) and boasting of the Cornish kibble, "the most popular of the improved hoisting machines in 1871"[3]. Most novel was their method of co-operative or lease-mining which they called "tribute pitch", by which a small group of miners would contract to work a portion of a mine on the share principle, meeting their own expenses and paying a percentage of their profits to the owners. For instance, "two Cornishmen rented the Dead Broke Lode, a branch of

1 Ibid., p. 254.
2 *Gulch of Gold.*, pp. 152-6.
3 Wallihan, p. 162.

the Gregory, and their ore, crushed at Charles Walker's mill by Joseph Westover, yielded ten ounces per cord"[1]. Lease-mining became standard practice, put the mines on a paying basis and promoted the reputation of the Cornish who "seldom joined unions and engaged in strikes, all of which won the gratitude and respect of their employers"[2].

One of the first Cousin Jacks to appear in contemporary accounts was the underground foreman at the Gregory Consolidated, Captain Rule, who in 1866 conducted Bayard Taylor, the correspondent of *The New York Tribune*, on a tour of the mine[3]. Rule arrived in the Gulch in 1863 and, according to Caroline Bancroft, had been a captain in the British army, resigning his commission in order to try his luck in America. Apparently he was a gifted musician and in 1876 at Central City produced *The Haymakers*[4]. But he was also the main prop of the Gregory Consolidated, where he was responsible for all the shaft sinking. By 1868 he had lowered the main shaft to a depth of 375 ft with a precision that commanded the admiration of all: "Capt. Rule, under whose oversight the mining has been done, is confident that the next level, which will be some 20 ft below the last, will be in the stratum of ore from which the Black Hawk Company, which adjoins them on the east, have taken 275,000 dollars during the past". There was no reason to doubt his judgment; he had installed a Cornish pump of 9 ins diameter; he was in complete charge; and the local feeling was, "it is safe to say it couldn't be in a better"[5].

Two other early arrivals of whom we know a little, though they do not appear in any newspaper, were John Truan from Redruth and William Trathan from St. Blazey. Truan staked his claim in "the beautiful valley of South Clear Creek which six months ago lay almost as God made it (but which) is now covered with palaces of boards"[6]. His wife with four babies followed later by train from Chicago to Cheyenne and then by stage to Denver where "she had to get a room in a frontier hotel with Indians and all kinds of ruffians about". The last part of the journey to Central City was along the frightening road from Idaho Springs which a reporter described as:[7]

One of the most extraordinary pieces of up and down hill road it has ever been my lot to pass over; on a twenty-four hundred pound coach, with about

1 *The Rocky Mountain News*, 12 November 1869.
2 Lynn Perrigo, *The Cornish Miners of Early Gilpin County*, pp. 93-4.
3 *The Denver Daily News*, 21 September 1866.
4 *Gulch of Gold*, p. 270.
5 *The Rocky Mountain News*, 13 February 1868.
6 Ibid., 2 July 1867.
7 Ibid., 16 June 1871.

three thousand pounds of human freight and baggage on board, and this great burden all rolled from an altitude of about eight thousand feet up, in a distance of three and a half miles more than two thousand feet higher, thence in three and a half miles more, down five hundred feet lower than the point of beginning.

At Central six more children were born, making ten in all, but they managed to save only seven, and Truan himself died early when his diseased lung collapsed. William Trathan lost his wife through typhoid almost as soon as they reached Central and had to send their children back to Cornwall to be cared for by his mother.

By 1870 there must have been a considerable army of Cousin Jacks in the mountains. George R. Mitchell is mentioned as taking out "splendid ore" from the Illinois Lode, one of the most valuable mines in Gilpin County, and receiving "from Professor Hill 1,000 dollars for thirteen tons of ore from the North Star mine"[1]. James Hoskins is killed at the bottom of the 150 ft shaft of the Rising Sun Lode: falling rock tore away his scalp, crushed his skull and pressed the hair down into the brain.[2] And two days after Christmas 1869, in a saloon on Ute Creek, John Levering, "otherwise known as Cousin Jack, was shot three times through the heart when he picked a quarrel with an Irishman"[3]. The pattern was familiar enough and disturbances were being continually reported about Irish miners "who got drunk and vented their spleen on Cornishmen"[4], for whom one reporter at least had nothing but praise:[5]

Since I came here I have seen quite a number of toil-worn miners step up to the counters of the different banking houses and deposit what looked to be bunches of paper, but which, when unfolded, revealed balls of beautiful retort gold, varying in size from that of a hen's egg up to half the size of a man's head. I am informed that the Cornishmen are the most persevering and successful class of miners here at the present time. The banks claim to receive larger bullion deposits from them than all others combined. Their habits of economy and their steady industrious disposition accounts for their prosperity.

As elsewhere they appear to have been an affluent minority. They continued to arrive in the 1870's from Lake Superior[6], while in the 1880 State and County elections it was calculated that the Cornish vote would fill more than half of the ballot boxes in favour of Colonel E. F. Osbiston, "a Cornishman by birth, a gentleman by culture and a

---

[1] Ibid., 12 March, 23 April 1868.
[2] Ibid., 31 August 1869.
[3] Ibid., 27 December 1869.
[4] Ibid., 22 January 1869.
[5] Ibid., 16 June 1871.
[6] Ibid., 14 July 1876.

miner by training and experience"[1]. In 1870 the population of Gilpin County stood at 5,490, of whom about 15% are listed as of English nationality[2]. S. S. Wallihan's *Rocky Mountain Directory and Colorado Gazeteer* for 1871 enumerates about fifty Cousin Jacks of Central City, living in the five boarding houses or with their families on Third, Eureka, Lawrence, Gregory and Packard streets. There were perhaps a couple of dozen more out at Nevadaville to the west of Central, at Black Hawk and at Idaho Springs on the South Clear Creek, eighteen miles from its junction with the Platte River.

But the prosperity they brought to the gulch towns was soon chipped away by swindlers and share-pushers, as Wallihan reports:

The dirty work is done by the capitalists themselves and their Fools. Old Bullion Bull of New York purchases valuable mining property from honest men in Colorado, who are compelled to sell their lodes from impecuniosity at half their real value. They sell out, Bull Bullion buys up, knowing that he need not do anything. Labour will be cheap, and he knows that there is wealth in the mine.

The Cousin Jacks, however, flatly refused to rate their labour cheaply. When in 1873 A. N. Rogers and George Randolph, the two representatives of eastern capital, declared that they must reduce the daily rate from 3 to 2.50 dollars because of rising costs, the Cousin Jacks roared their disapproval in such a manner that it was rumoured in Denver that "the Cornishmen had broken out in bold defiance of the law and that blood had been shed . . . and that the rebels had taken possession of the town and were intimidating the inhabitants". What really happened was quite different.

It was on the morning of 10 November that the news broke about wage reductions, whereupon the Cousin Jacks "threw down the implements of their labour", and for the rest of the day apparently did nothing but "talk and demur". In the evening they met at Turner's Hall in Central City under the leadership of Tom Turrell and were joined by those miners on contract wages who were not personally involved in the dispute. The next morning about 225 of them picketed the Leavitt, Briggs, Smith and Parnlee and the Bobtail mines, formed up in "line of procession" and marched to Black Hawk. They laid their complaints before Professor Hill and then returned to Central for another "secret" meeting. After their mid-day meal they strode out again, this time to capture the support of their uncommitted brothers at Nevadaville. But some hotheads, cloudy with liquor no doubt, clam-

1 Ibid., 1 February 1880.
2 Perrigo, p. 93.

oured to seize the mines by force and were only sobered when they saw the sheriff's posse arrive. By the evening calm had been restored and the third mass meeting at Turner's Hall passed off quietly[1].

The condition they laid down for a return to work seemed sensible: an acceptance of the wage reduction "if the mine owners will exert themselves to bring about a reduction in the price of ALL KINDS OF PRODUCE". Their case was a very strong one for the prices of groceries, hardware, dry goods and clothing were all inflated. A sack of flour cost almost twice as much in Central as it did in Golden, and so did fresh meat. "Marketmen", it was alleged, reaped as much as 200% profit. And they seemed to hold all the trump cards for since "they have the greater portion of money in the locality, they are liable to hold out indefinitely, or go into mining in other quarters". A long and bitter struggle seemed inevitable; eight Cousin Jacks were arrested on charges of conspiracy and fined 200 dollars each, and this so enraged their brothers at Caribou that they donated 150 dollars to a fighting fund. But harsh economic facts finally began to work against them, in spite of their claim that no other miners could take their place, not even Welsh coal miners. Lukewarm sympathisers at Nevadaville and Black Hawk defected; their own savings were soon finished; banks refused to extend credit; taxes went unpaid and their furniture was distrained in lieu; trade in the gulches was at a standstill; and so they lost all popular support for their case. Finally and most cuttingly of all, their best American friends bluntly told them to go back to Cornwall where they belonged: "The favourable circumstances which have enabled the Cornish miners to accumulate over a quarter of a million in the banks of this city, and many to return to England comfortably fixed for life, have been destroyed by this strike"[2].

The reproof was well-intended for the Cousin Jacks commanded much sympathy as they marched through the streets in a body. People for the first time took notice of their powder-blasted faces and realised how thin was the line that separated life from death, for though accidents here were not as frquent as on the Comstock, they were not less devastating. John Wall and James Hambley were both maimed when a Giant Powder cartridge exploded in their faces and filled their chests with powder, rock and sand[3]. On two occasions Richard Pearce was almost killed; in 1871 he fell 116 ft in the Fisk Lode, dislocated both

---

1 *The Rocky Mountain News*, 12 November 1873.
2 Ibid., 25 November 1873.
3 Ibid., 16 June 1874.

ankles, broke one leg near the hip, tore so many muscles that the surgeon feared that he would never be able to repair them all, and sustained injuries extensive enough to kill any ordinary man. After two years in hospital, he was stoping at the bottom of the Hunter when a loose rock fell on him, dislocating all the bones between the toes and ankles of both feet so that "his sufferings were intense"[1]. A year later there was yet another shocking accident with explosives, when John Rolfe, in a moment of forgetfulness, held a lighted candle and smoked his pipe over a large box of Giant Powder. His left arm was lacerated, one thigh gashed, one eye destroyed, and his left cheek laid bare to the bone. The body of his mate, James Penpraze, was picked up full of copper shell, the surgeons extracting over a hundred pieces[2]. But sympathy was not enough, for winter arrived bringing real privations. So each Cousin Jack decided for himself whether to return to work on reduced pay or to leave for Cornwall, and the strike was over.

Fifteen years elapsed before a recurrence of the trouble, a time when the Cousin Jacks picked the plums of the jobs and the Irish picked quarrels. In 1875 the Cornish are described as being "far in advance of the Americans", earning from 2 to 10 dollars a day and some even as much as 50[3]. One case has been quoted of six of them making a profit of 100,000 dollars in twenty months from leasing the Ophir, and not unnaturally returning at once to Cornwall[4]. The Irish troubles were generally no more than the appearance of "more sticking plaster than skin on their faces"[5], though there was one very ugly incident that made them all bury the hatchet for a time. One Sabbath August afternoon at a race meeting by a mountain lake, Cousin Jack John Williams, brother of the "notorious Dick the wrestler", and the Irishman Owen Ryan, both of them almost dead drunk, were struggling through an argument about religion. Ryan suddenly pulled a gun and shot Williams in the stomach. He in turn, staggering to his feet, picked up a rock and smashed Ryan's head "to a jelly". Not surprisingly, Ryan died instantly and Williams a few hours later. The latter can hardly have been comforted by the last words he heard from his wrestling brother: "Well, Johnny, you can afford to pass in your chips since you killed that son of a . . ."[6].

The Cornish-Irish feuding only came to an end in the early 1890's

1 Ibid., 20 January 1875.
2 Ibid., 27 January 1876.
3 Ibid., 3 July 1875.
4 *Gulch of Gold*, p. 256.
5 *The Rocky Mountain News*, 7 September 1875.
6 Ibid., 21 August 1877.

when their monopoly was threatened by the recruitment of cheap labour from Central Europe and Italy which provoked a serious confrontation with management. The first rumblings were heard in 1885 when *The Rocky Mountain News* of 13 June reported:

Miners, prospectors and mine owners in Wide Awake district have sat down on the Cornishmen, who form a large majority of the miners in Gilpin County. At a meeting Thursday of the miners and mine owners of Mellorsville, resolutions were passed and agreed to that no mine owner or miner in the Wide Awake district should employ Cornish miners or lot contract work or leases on mining property to Cornishmen.

In *The Gilpin County Observer*[1] the cause of the Cornish was championed by an attorney, W. C. Fullerton. He roundly accused A. N. Rogers, the manager of the Bobtail Consolidated, of deliberately attracting cheap foreign labour, and argued that the future position of the Cousin Jacks was abundantly clear: "What they want is their inalienable rights; the right to the just fruit of their labour; the right to be protected from all forms of aggression; this they will have or revolution". Was it therefore morally right to import "dagos" to work for as little as 1.75 dollars a day, "a class of miners whose only history is one of lawlessness and anarchism"? There had been some stabbings and shootings at the Egyptian mine which scared the Cornish[2], so Fullerton called upon them to close their ranks to repel the intruders, whom he described as "a horde of men who belong to the lowest class of the civilized world". Such violent comment might have passed unnoticed but for his demand that the Cornish should organize themselves into a labour union[3]. This was an anathema to the rival paper, *The Register-Call*, which reminded its readers that the Cornish, when they arrived in America, had not hesitated to underbid American labour to corner the best positions but now "would deny to others the rights and privileges extended to them". While this point of view was arguable, there seems no doubt that Fullerton had appointed himself the Cornish leader in order to climb on their backs to political office[4]. However, he rallied them to a meeting at Turner's Hall and persuaded them to agree to a declaration "That we will do all that lies in our power, without a violation of the law, to keep the wages of men who work in our mines up to the present standard and to the end we say the Dago must go".

1 *The Gilpin County Observer*, 4 January 1888.
2 Ibid., 11 January 1888.
3 Ibid., 18 January, 1888.
4 Perrigo, pp. 109-110.

Two Cousin Jacks had spoken out with unusual Celtic fervour. Isaac Williams took as his theme, "the injustice of men who had changed this place from a wilderness to a comfortable and prosperous mining camp, who had graded roads along the hillside with infinite toil, being driven out to make room for a cheaper and poorer class of citizens". Harry Trezise, however, tried to steer a middle course; he was sorry for the plight of the "dagos"; and he respected the position of the mine owners and the superintendents, especially Richard Rickard at the California. But the truth was that he could not support a wife and five children on 2.25 dollars a day[1]. Therefore, he argued, "this thing must be nipped in the bud, this tree must be pulled up by the roots before it has had time to take root". Then, as he wound himself up for the peroration, he launched into a "rendition of Tennyson's *Lady Godiva*", the wife of the rapacious Earl of Coventry who rode naked through the streets and so "took the tax away and made herself an everlasting name". Since no Cousin Jenny volunteered to emulate the Lady Godiva in the streets of Central City, the meeting had to be content with the appointment of a committee of five to draw up the constitution for a Miners' Union and to investigate a rumour that A. N. Rogers was paying his Tyrolean miners less than 2.25 dollars a day[1]. But the results disappointed them for the case against Rogers was never proved, a fact that "appeared to present the Dago question in a new light to many of the audience and had somewhat the effect of a cold shower bath".

The Cornish proved sensible enough to see that their long reign was over. They found that the Austrians, Italians and Serbs were competent and industrious, and some abandoned an old tradition by marrying their daughters; Ernest Semmens's bride in 1902 at Nevadaville's church was the handsome Marie Negri. Inevitably the Cornish pattern of Gilpin began to change, tea was supplanted by chianti, and the name over stores ended in "-ini", where once they had begun with "Tre-", "Pol-" and "Pen-". The most important change, however, was that the Cousin Jack no longer laboured in the mine but himself became an employer of the exiles from Europe.

Few could have regretted the passing of the silver age of the Cornish for the toll of human life had been heavy; fatalities like that of George Richards, decapitated by a blast in the Gregory[2], and James Lanyon, thrown from the bucket to the bottom of a shaft at Nevadaville, were

1 *The Gilpin County Observer*, 1 February 1888.
2 *The Rocky Mountain News*, 8 January 1876.

almost weekly news. But one August day in 1895 twelve men, Cousin Jacks and Tyrolese, were trapped by a wall of water in the Gregory Bobtail and only one survived. Henry Prisk and his teenage son scrambled up into a disused winze to escape the rising water, but the boy was overcome by the foul air, fell back into the torrent and was carried away; his father only avoided the same terrible fate by continuing his climb in the darkness until he found a ledge that miraculously offered him an air crack, and there he remained until found days later by the rescue team[1].

Others, of course, carried around inside them the germs that would eventually destroy them too. Such was William J. Vine of Denver who for years panted for his life as he tried to draw air through lungs perforated by dust and grit. He was only kept alive by medical inhalers that "made his feet to move and his fingers to tremble". In the four months before he died in 1958 and in between heart attacks he struggled to put pen to paper to write about the Cousin Jacks he knew in Central City who had risen to the ranks of superintendents. He wrote of Tom Dunstone at the Running Lode, Stephen Hoskin at the Carr, Herbert Bowden at the Calhoun, James Carlis at the Sleepy Hollow, William Bishop at the Cook, James Williams at the Alps and William Bailey at the Original; and such characters as Ed. Harris, the Cornish undertaker, who organised the annual Gilpin County picnics in July that became the rallying point for Cousin Jacks from all over America.

Vine was born in Black Hawk in 1885 and never saw Cornwall or his father, who was killed in the mine before he was born, as were three of his uncles, while four others died of silicosis before they were forty. Death was, it seemed, never far away. His grandmother, on her way to join her Camborne husband in Central City with three children, fell into the harbour at New York and was at first thought to be drowned (a mishap which she always referred to as "the first time she died"). When his father was killed and his mother married again, Vine was sent to live with her brother at Butte City, but he too was killed, and so he returned to Central to be drawn into the only kind of employment he knew. When he was eighteen he struck out on his own for Bisbee, "wild and woolly"; and then to Jerome, Tombstone and Naco on the Mexican border. Here the Arizona Rangers, "a tough bunch, good horsemen, fast with their guns", rode out to hunt down American and Mexican gun-slingers, bank robbers and cattle rustlers. Soon tiring of the heat and the blood of the fiestas, rodeos and bull-

1 *Gulch of Gold*, p. 234.

The most famous of all the iron mines in Michigan, the Chapin Mine at Iron Mountain, showing the engine-house that contained a Cornish pump.   [*Iron Mountain News*]

The Chapin Mine's enormous Cornish pump, which has now been moved to a permanent site for its preservation. At one time it was the most powerful in the United States.
[*Iron Mountain News*]

Cornish miners engaged in a drilling contest at Eldorado, Colorado, during the 4 July celebrations. [*Denver Public Library*]

The art of timbering in a mine at Central City, Colorado. Notice the Cornish-type candlesticks and shovels. [*Denver Public Library*]

St. Just-in-Penwith, Cornwall, looking towards Botallack and Levant mines and the Atlantic. It was from here that Richard Thomas and others emigrated during the severe slump in tin. [David Mills, St. Just]

Wardner, Idaho, where Richard Thomas from St. Just-in-Penwith tried to make a living in 1904. [Idaho Historical Society]

Rocky Bar, Idaho, as depicted in 1880, where William Rowe of Redruth was manager in a mining population of more than 5,000. In 1961 it had shrunk to one. The Cousin Jacks have stripped the hills of almost all the timber but today Nature has re-covered them with pine and removed most of the miner's works. (Idaho Historical Society)

fighting, he returned to Central City, disconsolately joined the ranks of the unemployed, and then made off for Butte to work at the Anaconda. His salvation was at the hands of his New York wife for "she had never lived in a mining town and was scared to death, seeing men die with miners' disease and hearing of them getting killed". She made him leave the mine and work with the Denver Tramway Company, a decision that few Cornish girls could have made but one that gave him a lease of forty years of life, even though the damage to his lungs had already been done.

Cousin Jacks, accepting their "consumption" as an industrial risk, were forced to pin their faith on useless quack medicines in bottles that bore the label "Sure Cure for Consumption". These can still be found on the litter dumps of Colorado's ghost towns. Scarcely any miner was immune and many arrived in Central already infected, like Semmens from Penzance who, to the eight of his children born there, added four more when he was fully employed at the Gregory Bobtail. His crowning achievement was to bring them all to adulthood at an altitude where the pneumonia of winter was an easy killer and where smallpox, scarlet fever and diphtheria could strike, whatever the season. But the 8,000 ft Rockies were too cruel for his diseased lungs. Almost too late he decided to move to the lower plain at Golden and there invest in a livery business of hacks, buggies, funeral carriages and ore wagons that would employ the energies of the entire family and made them independent of the mines.

After he had coughed himself to death in 1894, the business was sold and his widow invested the money in a Cornish boarding house, always a safe security for women with capital. For a widow left unprovided for, financial survival depended on taking in the dirty linen of others or turning her hands to millinery or waiting at table in the boarding house. The hours were long and the work exacting, but there was always a time set apart for church and chapel affairs, for choir meetings and the literary evening, when she would set herself to read the Bible, a newspaper or some periodical, living for the day when she might see her grandchildren entering upon their educational inheritance far sooner than those left behind in Cornwall.

Thomas H. Williams from Phillack, conditioned to the crushing burden of labouring in the mines from the age of nine, and married to Margaret Ann Floyd with two infant girls to feed, was faced with the hard fact of separation when he was twenty-five. He reached Upper Michigan only to find the miners taking the trails to Colorado and so joined them, making his way as best he could: by sailing-boat across Lake Michigan, then by devious routes to Leavenworth in Kansas, and from there to Denver, a distance of 650 miles through country where Indians were particularly active. He recalled that no one was allowed to board the stage coach unless armed; that passengers were instructed to overturn the coach in the event of an attack; and that stage traffic was dislocated for several days because Indians had burned a relay station, killed the station hands and stolen all the horses. Once at Denver, Williams lost no time in heading for Caribou, a new camp at an altitude of 11,000 ft, where he was able to earn enough to send for his wife and children. It was a forbidding and isolated outpost and women were scarce. There were only four and they were forced to act as general nurses. Three of them came from Cornwall: his wife, one "Granny" Moyle and a Mrs. William Nichols. But Williams was too good a miner to stay there long, for he found the veins too "pockety" to be economic; so he roved out to Central City, where he leased several properties and ultimately rose to the position of superintendent of the Gregory Bobtail. This was a satisfying achievement after a lifetime of trying to make ends meet in camps so remote that water from the streams cost a dollar a barrel and a head of cabbage five dollars. But the losses could be heart-breaking; four of their five children died of diphtheria; and Williams at fifty-four of silicosis.

Mary Jane Waters of Golden came from St. Just, the thirteenth or fourteenth (she could not remember which) of a quiverful of twelve boys and three girls, five of whom died in infancy. Her timberman father died when she was six and every year or so came the tidings of brothers dying in Africa or America. At the age of sixteen she was alone with her mother and her youngest brother. The only other survivors of that large family was another brother at Nevadaville who paid for their passage, and they made the exhausting journey to his home there. Within two weeks of their arrival her younger brother slipped away to die, his frail frame weakened by the travelling and broken by the altitude. At last she found some security by marrying a Cousin Jack and by the time she was twenty-eight, no doubt prematurely old, had borne him seven children.

The wanderings of the Hoskins from the golden surf-beaten beaches of Perranporth began when Edward, the blacksmith, married Thomasin George. In 1874 with their three children they were emigrant passengers on board the *Circassian* outward bound for "the Basin of Mines" near "the early home of Evangeline" in Nova Scotia. At first they tried their hand at managing a boarding house there, and later joined a brother at New Diggings in Wisconsin, travelling along the trail from

Galena, already clearly marked out by the pyramids of boulders left there by the earlier generation of Cornish lead miners. Blacksmiths are less circumscribed than miners in the pursuit of their techniques. When his wife complained about the discomforts of the altitude near Idaho Springs and ordered the family down to Denver, he faced no problem in finding employment, first with the Colorado Ironworks and then with the Burlington Railroad. As it happened, the break-away from mining coincided with the building of the Burlington's short-cut into Denver and Hoskin took advantage of the cheap land and farms that could be bought along its route, homesteading on the prairie at Holdrege in Nebraska. In 1888 even better prospects opened up before him with the coming of the Rock Island Railroad; and so he moved to Beloit, a town that had begun to boom when it was known that the railroad would pass through it. Unfortunately the line was surveyed to pass eight miles to the north and Beloit passed into oblivion overnight, leaving Hoskin with two ranches of 400 acres each and no supply lines. But he refused to leave, hung on to his ranches with his two Cornish sons in control and moved his blacksmith's shop to Burlington.

Their children belonged to the new breed of Americans, even the two born in a cob cottage in Perranporth rising to positions of importance that would have been impossible in Cornwall. One was elected Sheriff of Kit Carson County for four years and Democratic candidate for the Colorado Legislature. The other, Henry George, whose education began in a one-room school in Perranporth, was continued in the make-shift school-houses of the mining camps and was supplemented by "an unceasing programme of self-improvement"[1], managed to combine a complete range of successful activities: farmer, teacher, founder of the family business known as the Kit Carson Abstract Company, Town Clerk, Treasurer and Mayor of Burlington, County Chairman of the Republican party and twice a Congressman in the State Legislature.

Beyond the festival town of Central City, for such it is now, in the remoter fastnesses of the Rockies sprang many other camps that the Cousin Jacks penetrated with their drills and picks, labouring at altitudes which seem to dwarf Denver to the level of the sea. Sixty miles from the border of New Mexico on the San Miguel River they were pinned down in their tents by Mount Sneffels and Wilson Peak, both reaching skywards to more than 14,000 ft. Telluride was the

---

1 *The Burlington County Record*, 6 January 1945.

camp but it could only be reached over the 10,000 ft Lizard Head Pass, so they only ventured there on special occasions like Christmas and Independence Day and when one of their clan needed the decencies of a Christian burial. Twenty miles from Telluride as the eagle flies over the 11,000 ft Red Mountain Pass, and nearly 100 miles by mountain railroad, was Silverton. Prospectors needed to be hardy adventurers to stake out their lodes here in the desolate gulches of the San Juan Mountains, for deep snow made them treacherous for most of the year. Even after a treaty with the Ute Indians had opened up a track to Silverton from the camps in northern Colorado, the difficulties of access were immense, the machinery for the smelter and the silver bullion being shipped on burros from Colorado Springs or to Pueblo. Yet here at the Philadelphia, the Pride of the West and other mines a clutch of Cousin Jacks braved the hurricane winds and the drifting snow: Ernest Allen, three Barlings, John Bowden, Ambrose Bray, John and Jim Curnow, Ben and Joe Dunstan, Frank and Bill Henwood, John Hosking, Bill Hocking, James and John Kitt, Matthew Moyle, John and Jim Peneluna, John and Walter Prout, John Prisk, Ben Toy and Jack Paul, who was an assessor for San Juan County. Today they are remembered by Mrs. N. C. Maxwell, Maud Pearce as she was when she left Pengegon near Camborne to join her father in 1884 at Silverton, who believes that she is now the only Cornishwoman living in the region.

Another difficult camp to reach was Caribou at the foot of snow-covered Arapahoe Peak to the north-west of Central City. So named by George Lytle who had prospected in the Caribou district of British Columbia, it was Colorado's first paying silver mine and by 1875 had attracted to its desolate mountain side about 3,000 hungry fortune-hunters. Here, it has been said, the winds were born and here a winter's day could be a whole year long[1]. Few could endure the deadly blizzards without some protection over the face for breathing; to walk meant crawling on hands and knees; and they had even been known to put an end to a Cornish July picnic. The Cousin Jacks and their families took a fearful battering, not only from the onslaughts of winter, but from the summer heat, when typhoid, dysentery, smallpox, cholera, scarlet fever and cerebro-spinal meningitis would take their toll of young lives. Ghost town now, for it never really recovered from the devalua-

[1] See John and Doris Buchanan, *The Story of Ghost Town Caribou* (Boulder: 1957): and David H. Stratton, 'The Cousin Jacks of Caribou', *The Colorado Quarterly* (Spring 1953), and 'The Rise and Decline of Caribou', *The Colorado Magazine* (Vol. XXX, No. 2, April 1953).

tion of silver in 1893 and a blizzard of 1920, Caribou is a memorial to the mining pioneers who moved on and to those of their children they left behind in the ground of a wasteland whose only concession to a vanished population is the weathered boards that bear their names: W. M. Hicks, died 1875, age two; Emma George, died 7 August 1877, age nine; and the three children of the Richard family, ages ten, eight and three, wiped out in a single week in July 1879.

Along yet another of the silver spokes that radiate from Central City is Georgetown, set in motion in 1864 by the discovery of the Belmont Lode and other camps that exploded around it: Idaho, Spanish Bar, Red Elephant and Silver Plume. Conditions here in 1869, according to a newspaper report of 17 January, amounted almost to a reign of terror:

Between the hours of ten and eleven o'clock on Sunday night a drunken and besotted crowd, under the leadership of Mike Royce, undoubtedly one of the worst characters in this section, and a man of the name of Ward, who had but just recovered from the effects of a pistol shot wound which he had received in a house of prostitution in Central, went to the house of a Cornishman named Arthur, occupied by himself, his wife and a young woman, and demanded an entrance in two and a half minutes, or they would burst the door. During this time, two shots were fired through one of the windows of the house by the parties outside. The occupants had retired but Mrs. Arthur got up and opened the door, the crowd entered and immediately commenced firing their revolvers into and through the ceiling, one ball embedding itself into a slat of the bedstead upon which the young woman was sleeping. These heartless wretches destroyed every article of furniture and household goods on which they could lay their hands, took the stove out of the front room and brought in chips, piling them around the box on which the stove had stood, and set fire to them with the object of burning down the house. But the crowning act of infamy and savage barbarity was perpetrated by these brutal wretches when they seized the young woman, who was attempting to escape from the house, and compelled her to stand in a corner of the room and fired several shots at a mark about six inches above her head.

Ten years later, however, as more Cornish flocked to the town, the rowdy element seems to have disappeared and Cousin Jack was being described as "a good boy" and "most deserving of praise"[1]. So, to the boulder-strewn gulches came John Henry Vivian of Camborne in 1875. He worked the lodes for five years, sent home for his wife and children, and saw his sons venture to mine in Globe, Arizona, and enlist in the American Expeditionary Force to France in 1917. A Penberthy

---

[1] *The Georgetown Colorado Miner*, 2 August 1879.

managed a boarding house there in 1880;[1] and there soon appeared the far-famed firm of Edward Eddy, Hoskins and Company[2]. Just west of Georgetown and Silver Plume stood the less romantically named Brownsville, thriving none the less on a school, a church, a boarding house, a "lyceum", a brewery, the usual complement of saloons and a ring for Cornish-style wrestling. It died suddenly when the dump of the Seven-Thirty mine, high on the mountain side, slithered down its smooth slope and buried the entire settlement. Here that Cornish wanderer, Stephen Thomas, had once led into musical battle the Terrible Silver Cornet Band, so called after a mine of that name; and here took place on a memorable evening in August 1877 one of the longest wrestling bouts ever recorded, when George Wedge and John Hall fought each other over a distance of four hours and eleven minutes[3].

An exceptionally successful Cousin Jack was Edward Eddy, who had received some training at the famous Royal School of Mines in London. Born in 1840, he was thirty-one when he entered George-town with 15 dollars in his pocket, ten of which he characteristically remitted to his wife in Cornwall; then "with his blankets on his shoulder", he searched for work. Taken on by the Welsh superintendent of the East Terrible mine, he made so good an impression that he and William H. James went into partnership. Within a year he had saved 600 dollars and bought several leases, turned these into profitable concerns and begun the building of the Silver Plume Sampling and Concentrating Works. In 1878, anticipating a decline of business at Georgetown, he transferred his milling to the new and sensational carbonate camp at Leadville, where he established a virtual monopoly in the buying of ores. At the end of their first year of business, James and Eddy milled silver and lead to the value of nearly 750,000 dollars. The arrival in Haw Tabor's wild town of the Welshman who became its Mayor and of the Cousin Jack, "untiring in effort and liberal in dealing with mine owners", was a triumph for Celtic industry, "vim and energy". Their extensive smelting works, covering 8,000 sq. ft. and in-cluding in its machinery "two sets of Cornish rolls", were completed and blown in on 3 July 1878. How mammoth were the dealings of these two Celtic business executives so far from home may be gauged from their ore and bullion purchases for 1879, which amounted to 3,356,845 dollars. Agent for the Omaha Smelting and Refining Company, and

---

1 *The Colorado State Business Directory* (1879).
2 Ibid (1869).
3 *The Colorado Miner*, 4 August 1877.

Mather and Geist of Pueblo as well, Eddy in 1879 could well afford to return to Cornwall "to visit his aged father" and to look upon "familiar faces and the scenes of his early life"[1].

Eddy's Leadville was almost the last of the wild-cat camps. At the distressing height of 11,000 ft, within twelve months nearly 10,000 fortune-hunters poured into its ravines and gulches. They pitched their bones in draughty tents on frozen ground or in cheap hotels, where the beds were occupied for twenty-four hours a day on an eight-hour rota, or at the Mammoth Sleeping Palace with its double tiers of bunks that slept 500 pairs of eyes a night at fifty cents a time. B. C. Keeler, *The Chicago Times* correspondent, was terrified of the place when he visited it in January 1879, not to mention the hair-raising ride on the train and stage by the edge of falling precipices. He found Leadville a camp of only six streets, on which jostled 6,000 wanderers, of whom 300 were women, 200 of them married, and 100 who ought to have been. The prostitute was held in high regard for the decorum of her cabin where "there is nothing gross done, seen or said". In conversation you could never be sure whether you were talking about a lady of easy virtue or a mine, for the same names did for both: Henrietta, Del Monte or Little Eva. A pair of accurate Navy revolvers was advised as a necessity. Everywhere abounded men and women sick with lung and sinus afflictions, who were sure to die as soon as they faced two fresh hazards, poisoning from the lead in the streams they drank and from the bismuth fumes they breathed from the belching smelters. Almost everyone coughed himself into a coma and nursed his aching joints in frail wooden unplastered cabins that gave no protection againt the bitter winds. Night after night the miners slept cold and, after the warming coils of the whisky had slunk away, became delirious with lung fever. Too few blankets and too much liquor accounted for as many as ten deaths a week.

Leadville was hardly a Cousin Jack's New Jerusalem and perhaps it is not surprising therefore that only a few of them can be discerned in a Leadville directory of 1880: Anthony Barrett, proprietor of the forbiddingly named Carbonate House; Mrs. A. Barrett, dressmaker; William and Albert Chellew; and Frank Mitchell, Philip Oates, John Odgers, Alfred Pierce, Alfred Pomeroy, David Snell, George Trevillian and F. Treworjy. For some, however, it was still "the place" to settle, even after the repeal of the Sherman Silver Act which sent the bonanza tycoons scuttling to close their banking accounts and brought

---

[1] L. A. Kent, *Leadville in Your Pocket* (Denver: 1880), pp. 110-11, 177-8.

about the ruin of some who had hoped for little more than a quiet retirement on their moderate savings. No doubt they agreed with one observer[1] that Leadville had been "vilely represented", that the mountain air was "wonderfully bracing", the summers wholly delightful and the winters no more severe than in Nebraska or Iowa. Among them was Lavinia Johns who in 1888 at Eureka, Nevada, came across William Spargo from Stithians and without more ado married him. Together they wandered throughout the westlands and then settled for a restaurant in Helston. But this was no life for a hard-rock miner, and Spargo was soon on his way again to California. After two years his family rejoined him but almost immediately he was off again to Leadville. From here his wife resolutely refused to move again, so he bought her a house at the top of a mountain to which he would return unconcernedly after bouts of absence in Arizona or Montana; and here in the thin air far above the timber line he and his wife chose to be buried.

For those left to roam the Colorado ranges there remained one more camp to be exploited for the elusive fortune: Cripple Creek. Situated about twenty-five miles to the west of Colorado Springs, it was the home of rich cattle ranchers until the cowhand "Crazy Bob" Womack struck the El Paso lode and exposed the gold-bearing rock that lay beneath fifteen miles of lush meadows, soon to be stripped of their grass by more than 3,000 exhausted miners. Their hopes, however, were to be dashed for hardly had they driven their shafts down the Portland, the Pactolus, the Pay Rock, the Plymouth Rock and the Pinto mines than the financial panic of 1893 drove owners to cut their costs and demand that they should work nine hours a day for 3 dollars. The strikes that followed were the most violent ever witnessed in the mining industry and were symptomatic of the industrial malaise that swept the whole country and manifested itself in the Pullman strike and the march of Coxey's "army" on Washington. Desperate and angry miners dynamited the hills around Cripple Creek so that they could be blown skyhigh at the touch of a plunger, made their own guns to fire beer bottles filled with dynamite, and commandeered a train, which they drove half-way across Kansas before being stopped, to take them to join Coxey's "army"[2]. But the reports of the disturbances give no indication that the Cornish were involved; presumably, if they found conditions too unacceptable they simply moved out.

---

[1] J. L. Loomis, *Leadville Colorado* (Colorado Springs: 1879).
[2] See B. M. Rastall, 'The Cripple Creek Strike of 1893', *Colorado College Studies* (Vol. II, June 1905); and Mabel Barbara Lee, *Cripple Creek Days* (Doubleday: 1954).

One who found Cripple Creek to his liking was Samuel Rowe. Born in a damp cob cottage on the rain-sodden moors between St. Cleer and Pensilva, he watched nine brothers carried to their graves with consumption and then himself escaped to Michigan, his left hand mangled and mutilated from an accident in a Cornish mine. After four years in Iron Mountain he returned to St. Austell to claim his bride, with no desire ever to set foot again in Cornwall, though fate was to make him decide otherwise. Six years and four children between 1884 and 1890 determined him to exchange the frosts of Michigan for the heat of Arizona, but Bisbee was gripped by a typhoid epidemic and his wife died within a day or so of their arrival. Without hesitation he evacuated his four children, sick though they were, and bundled them on to the first train out of Bisbee, his intention being to place them with relatives in Cornwall. The youngest was only a year old and could not have survived the journey but for the father searching for fresh milk at every train stop. At Deming in New Mexico, driven to the very limits of distraction, he had to take them off the train until they had completely recovered. Then followed the long rail haul to New York, the sea passage to England and the long awaited meeting with his sister. She had offered to mother the children, for she too had recently sorrowed in Silver Plume where her husband had been killed. Rowe, however, had to return to America to provide for them and this time plotted a course for Cripple Creek. He had read of the astounding successes of entrepreneurs like W. H. Stratton, Charles Tutt and Spencer Penrose who, descended from ancestors in Bristol and Cornwall, directed an empire of metals that stretched from Arizona to Alaska which would one day be the patron of medicine and education throughout Colorado. To his consternation Cripple Creek in 1896 was a town on fire but he decided to stay, living in a tent until it had been rebuilt. The decision proved to be right for, after leasing mines when he was "flush" and working for companies when he was "broke", he was in the permanent employ of the Portland Mining Company to prospect on the Rio Grande around Pecos in New Mexico. Finally he returned to Cripple Creek to invest in property and to invite his Cornish children to join him after some twenty years' absence. But only one did, so he married again (a Cornish widow) and lies buried now in Colorado Springs among the trailblazers and pioneers. Raised on high among the pines and tumbling waterfalls of the granite masses of the Rockies, he hardly had to worry about the decline and fall of Cripple Creek, unlike W. T. Henwood and others who were then compelled to seek work on the electric tramways of Boulder. Nor was he too closely tied to the land of his birth like some whose only desire was to be washed back again to Cornwall's coasts for burial in the same earth as the warriors and chieftains of an earlier age of iron and bronze. David Dingle, for instance, from the bleakness of Callington, trailed through Indian country to Silver Plume, lost the sight of one eye and returned to Cornwall. His brothers too, the world their oyster in South Africa, South Dakota and Colorado, came back to be buried in Gunnislake after arranging for an embalmer to ship them home. And to many of their descendants left in Colorado, "home", in a strange unaccountable way, still means Cornwall. As Mrs. Mary Risch of Ouray says: "It is strange how close I have always felt to Cornwall; in fact, though I was born in Nevadaville, and have spent all but three years of my life in this country, those being spent at St. Just, I usually feel more Cornish than American".

A man who knew the Cousin Jacks at close quarters in Colorado was T. A. Rickard, the Cornish mining engineer who was himself descended from the famous Porthtowan family. His grandfather was probably the first accredited miner in California, his father was connected with mining ventures in Italy, Switzerland and Russia, and at one time no less than eight members of the family belonged to the American Institution of Mining Engineers. T. A. Rickard was perhaps one of the world's greatest engineers, whose mining experience and scholarship can be sampled in *The Mining Magazine*, which he founded, and *The Engineering and Mining Journal*, which he edited. But his apprenticeship was served in Colorado; in 1885 he was his uncle's assayer in Idaho Springs, where the superintendent was John Curnow; in 1886 he managed at Central City his uncle's mines, the Kansas, the Kent County and the Californian; and then he took charge of more mines in Leadville. After some years at the Union mine in the Calaveras County of California (where his emoluments included 300 dollars a month, a house, Chinese servants, a horse and a Cornish book-keeper, John Roscrow), he returned to Colorado to manage the Yankee Girl at Red Mountain and the Enterprise at Rico, where he introduced contract work on the Cornish plan. It was at Denver that he married in 1898 and it was from the capital of the Centennial State that he departed to investigate a host of mines in Arizona, Australia and Alaska.

Having worked alongside miners from every conceivable nationality the world over, this Cornishman may perhaps be forgiven for asserting, "the men that I liked best were the Cornish miners". And had he seen them spilling their blood a generation earlier in Nevada, it is certain that he would not have altered his opinion.

VI

HUNTING SILVER TO THE DEATH IN NEVADA

In contrast to the mines of Michigan and Colorado, those along the silver lodes of Nevada imposed unbelievably harsh strains. Often more than 2,000 ft down, where the temperatures ranged from 100°F to 120°F and "where the water comes hissing hot from Tartarus", they exacted a terrible toll of one accident a day the year round, of which one in five was fatal. As one observer reminded those thinking of emigrating to the Comstock: "Ten thousand tons of rock may fall upon you and leave not the semblance of humanity in your manly form; you may strike water in a drift and your lifeless body be found weeks after when the pump has again drained the mine . . . all these and a thousand more things may occur and for incurring them, you may, when old and experienced, get four dollars a day"[1]. A few, a very few, stumbled across a fortune, like the Cornishman Tom Jenkins who in 1859 made 20,000 dollars in a single day "by right of discovery"[2]. Thirty years later Dave Rowe, whose house sat on an old dump of the Ophir mine, was digging out a cellar when he struck a mass of ore. Knowing that it rightly belonged to the company, he mined it secretly, hauled it away under the cover of night down Six-Mile Canyon and reduced it in a private mill. But he was unable to conceal from the company the rise in his personal standard of living and was prosecuted; he thereby lost a fortune of 90,000 dollars[3]. But such fabulous finds were rare, and all too often a Cousin Jack's mining fame ended with his mangled body at the bottom of a shaft.

He knew no other Nevada than that of the desert, wilderness and mountain where the minerals occur. He could hardly be expected to stop at the old Pony Express station of Walley's Hot Springs and marvel how audible was the silence that carried to his ears the click of two stones brought together a hundred yards away. He knew nothing of the Nevada of lakes, streams, rich pastures and farms; and he could afford no more than a passing glance for that Indian pearl, 6,000 ft in the sky, Lake Tahoe, over whose smooth water glided the

1 *The Virginia City Territorial Enterprise*, 19 August 1875.
2 Ed. Duncan Enrich, *The Comstock Bonanza* (New York: 1950), p. 13.
3 Wells Drury, *An Editor on the Comstock Lode* (New York: 1936), p. 288; and Grant H. Smith, 'The History of the Comstock Lode 1850-1920', *The University of Nevada Bulletin* (July 1943), p. 215.

steamships *Meteor* and *Tahoe*, captained by Edmund Hunkin from the fishing village of Mevagissey in his own Cornwall[1]. As fast as his dust-caked boots could take him, he was bound for Virginia City in the fanged Sierras where, on a barren mountain side named after the sun, the silver had cropped out in a waste as unpromising as the Mojave Desert that spawned Calico.

Whether by car today or on foot a century ago, Virginia City never seems more than an anticipation, locked in as it is by the barren mountains that guard its approaches. The road twists and writhes in its ascent through pock-marked canyons where only an occasional mesquite serves as a landmark, much as when the first strikes were made in October 1859: "The ground was torn up in all directions with shallow cuts and pits; diminutive adits pierced the hillsides like the holes of sand swallows in a mound and the gray carpet of the sage brush was buried under unsightly heaps of sand and crumbling rocks"[2]. Today, however, in this blistering wilderness only the lizards scuttle for shelter beneath rocks ice-cold or burning according to the season. The miners have long since departed from the spiralling track that once linked Silver City to Gold Hill, at one time almost entirely Cornish towns. At Gold Hill hardly the shell of a single building remains; sign boards alone mark the sites of the Methodist Chapel, the brewery, the Oddfellows' Hall, and the Silver Star Lodge, the first of the Masonic Order in Nevada, burnt to the ground as recently as 1928. Silver City suffered the same fate in 1885 and today all that is left of it is a greying paint-peeling brick building, and heaps of rubble and ash where once stood the Miners' Hall and the Episcopal Church. The visitor disconsolately passes trees of thorn that snatch and tear and mounds of stones and red charred earth; a buckled buckboard beneath yellow tailings where an intricate lattice-work of hoists still probes the lode; the parched brown timbers of the cabins of brave prospectors; the redbrick Gold Hill Bar and Hotel that is now a museum; and then quite suddenly he is walking along the famous "C" street of Virginia amid the camera-clicking of tourists, the jingle of silver dollars and the metallic ring of gaming machines. In this tourist "attraction", complete with planked sidewalks and hitching posts, he cannot miss the breath-taking surprise of St. Mary's-in-the-Mountains but he almost certainly will not see Boot Hill a mile away. It encompasses no less than eleven cemeteries, grim monuments now of the ground needed to cope with the output of dead from the shadow of the mountain which gave a

1   See Richard J. Rowe, *The Story of Tahoe* (no date).
2   Eliot Lord, *Comstock Mining and Miners* (Berkeley: 1959, reprint of 1883 edition).

living and so avidly demanded life. Wooden fences, once clean and white but now warped and split by wind and sun, rail-in the graves of Cousin Jacks who dug out the Big Bonanza that the mint at Carson City changed into millions of silver dollars: W. H. Trevillion, Matthew Rapson, Richard Rogers and John Retallack "killed in the Union Shaft on the 26th of October 1879, aged 32".

Silver was not precious there until gold became scarce. The first mining immigrants, working down the Carson Valley on their way to California, were looking for gold rather than silver. So were a party of Mormons in Gold Canyon and other restless prospectors from the Mother Lode like the two brothers Allen and Hosea Grosch, and a shiftless sheepherder, Henry P. Comstock, otherwise known as "old Pancake". Searching for gold on the flanks of the Sun Mountain and in the bare gulches of Six-Mile Canyon they did not comprehend that the blue clay they threw away was in fact silver. A Judge Walsh of Grass Valley appears to have been the first to appreciate that the lode, named after the most feckless of grubstakers, was in fact a silver bonanza. He inspected the mines at Gold Hill, took a ton of the ore to San Francisco which assayed at between 900 and 5,000 dollars and returned to Washoe to erect machinery for concentrated extraction. When the news broke, adventurers at once swarmed around their new prey in spite of warnings that water was in short supply and that there was no timber within twenty miles. In the fall of 1859 they hopefully pitched their flimsy tents and watched them blown away by the Washoe "zephyr" as if in anger that the mountain to the sun had been re-christened Davidson after Rothchild's moneyed representative in San Francisco. But they were back again with the melting of the snows in the following spring. The rush to Washoe had begun, gathering momentum as it moved towards the Sierras and sustained by a new spirit of feverish excitement and romance associated with memories of the Spanish Main. Never before had bars of silver been seen in the streets of San Francisco. Merchants, clerks, stevedores, sailors, ranchers, fruit farmers and the riff-raff of the waterfront boarded the boats for Sacramento and jammed the trail to Placerville and beyond. As one writer has graphically described it:[1] "The motley train stretched in a broken line from Sacramento to the mines—old and young, waifs from many nations, who had drifted during ten years to the California goldfields, with every variety of dress and equipment, mounted and on foot, driving pack mules, burros and wagons (while) at the little stations on the route Piedmontese and

---

[1] Eliot Lord, *The Drama of Virginia City* (Reno: 1926), p. 37.

Cornishmen, Jews and Catholics, mechanics and scholars, honest men and rogues, snatched their food hastily from the same rude board . . . and slept at night on the same bed of straw".

Unwashed, uncombed and bitten by fleas, these Cornishmen made their tedious way to the valley of the Carson only to find their diggings ravaged by incensed Pah-Ute Indians on whose land they trespassed. Many abandoned their claims and returned to California. Business was at a complete standstill, and Virginia City, so named by the prospector "Old Virginny" in one of his drinking bouts, lay almost deserted and bankrupt, slumped across the richest ledge of silver in the world. By summer however the Sacramento road was again choked with miners and their burros. Everyone boasted their wealth at least in claims, and prices rocketed so high that a cabin, 10 ft square with canvas sides and muslin windows, cost as much as 500 dollars[1]. The smooth stream of silver from the new steam mills poured out in such a flood that on the historic road over the 7,385 ft high Echo Summit 950 mule trains were operating. Every mule was treated as if it were the most refined Arab horse, for no animal could better endure the altitude or pull such heavy loads over the most excruciating roads imaginable that only just managed to cling to the very edge of precipices. It was a road fraught with every kind of natural hazard but in its time the main artery that fed San Francisco with the silver its smelters needed.

Virginia City in 1862 still had a population of less than 4,000 but it included an important proportion of Cornish working the 500 claims that had been registered by over fifty corporations with assets that ranged from the 2,000,000 dollar Mount Davidson Company down to the mere 50,000 dollar Scoria Company. Wells J. Kelly's *Directory of Nevada Territory in 1862 and 1863* lists the following miners who appear to be of Cornish origin: Ebenezer Basset, George Berry, William Eddie, Joseph and R. Hancock, Alfred Jenkin, M. Mitchell, James Pope, George Roe, John Rowe and John Spargo. Edward Stephens is employed "at the Mexican claim", and Nicholas Treweek by the Sugar Loaf Quartz Crushing Company, while William H. Rouse is noted as being the superintendent at the Monte Cristo claim. Gold Hill was also beginning to echo with Cornish names: Stephen Blight, R. B. Buckley, John Craze, Benjamin Glasson, James Jenkins, William Luke, R. H. Mitchell, I. W. Richards and Nathan Winn. John Rule was part owner of the Marysville mill and Edward Rule the chief

---

[1] Ina Powers Sample, *The Miners' Movement for Statehood in Nevada* (unpublished M.A. thesis, Berkeley, 1932).

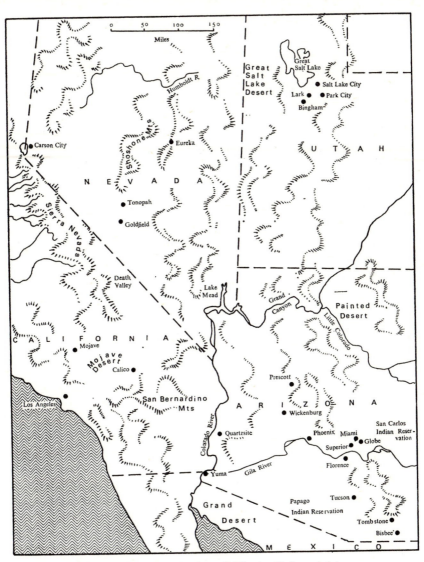

Cornish settlements in Southern California, Utah, and Arizona

engineer, keeping watch on the nine heavy stamps, each weighing 850 lbs, made in the Vulcan Foundry at San Francisco. John Tregloan was the superintendent of the Swansea Mill and Mining Company's mill in Gold Canyon, in which he and S. M. Johns each owned a quarter share; he was also part-owner of the Franklin mill, where ten men were supervised by William Johns. It was built in Grass Valley, cost 60,000 dollars, worked ten stamps, crushed fifteen tons of ore a day, and was driven by water brought from the Carson River along wooden flumes.

On the west bank of the Carson at Zephyr Flat, William S. Rowe was in charge of yet another mill. This one was also worked by water, run from a solid stone dam through a race 15 ft wide for a distance of half a mile. Over at the Potosi C. C. Thomas superintended his Cornish foremen, as did John Pascoe at the Central, so it is not surprising that with captains of industry such as these the Cousin Jacks found little difficulty in rustling jobs, whatever their capabilities. Altogether there must have been at least 100 of them scattered around the shafts and dumps by the end of 1863.

The life they had to lead at Virginia City was not very much different from that encountered at other camps: drinking saloons, Chinese bending over wash-tubs and cooking stoves, down-at-heel Indians, and the rows of white cabins "with gaudily furnished rooms, at whose windows the inmates sat waiting like spiders for flies"[1]. Their own houses were adequate by Mother Lode standards, being made of wood and comfortably furnished. If they boarded at a hotel, the cost might run from 8 to 10 dollars a week, but this was not exorbitant, for the food was usually excellent. They could well afford it, since a mill-hand could earn 3 dollars a day and an engineer as much as 8[2]. The chief problem was wage stability, and this the miner was determined to have as some compensation for the unusual dangers he incurred.

The first serious effort to ensure a standard wage of 4 dollars a day along the entire Comstock came in May 1863 when about 400 miners at Virginia City formed the Miners' Protection Association. The Cornish, however, do not appear to have been very active in promoting its aims, for their own pay was hardly ever in jeopardy. An F. Polkinghorn was elected to the finance committee in 1871 but he is the only one. Indeed, when in 1865 wages fell from 4 to 3.50 dollars at the Uncle Sam mine, it was the Cornish foreman who was blamed. The infuriated miners seized John Trembath at one of the lower levels, trussed him up with a

1   Lord, *Comstock Mining and Miners*, p. 98.
2   V.C.T.E., 12 June 1871.

coil of rope until he looked like an Egyptian mummy, lashed him to the main hoisting cable, whisked him up and down the shaft several times, and then left him on the surface, humiliated and almost unconscious, with a label tied to his jacket which read: "Dump this pile of waste dirt from Cornwall"[1]. So unkindly a demonstration can only be explained perhaps by the severity of the depression. Most miners realised how often they owed their lives to the Cornish who, when the worst dangers were from flooding, moved through the drifts with the utmost caution and care and by-passed the partitions of clay which they alone knew to be the natural bulkheads for water[2]. Twelve pumps coped with the water at the principal shafts of the lode but, until Adolph Sutro in 1865 began his famous tunnel between Webber and Corral canyons, it was the Cousin Jacks who showed the less experienced how to minimize the risks.

Wage stability was impossible to achieve, for booms and slumps came and went with the regularity of the seasons, and the miners suffered accordingly. In 1863 the market value of Comstock stock stood as high as 40,000,000 dollars, but in the next two years it fell to a tenth of that value. Buetween 1864 and the end of 1865 more than 10,000 miners drifted away to California, Idaho and Montana, victims of the cutthroat competition between the far too numerous operating companies[3]. But they were back again in 1866, though ill-prepared to withstand what was to prove the worst winter of their lives. The air was so cold and fuel so scarce that wooden boxes from California were worth more than their contents, women burned their corsets and cripples were warned to keep a special watch on their crutches[4]. The Sun Mountain, now the Snow Mountain, seemed the devil incarnate, starting slides, crushing cabins like eggshells, smothering men in their sleep, blocking the passes and holding up the stages so that, when the spring arrived, five tons of mail were waiting to be distributed. At six o'clock one February morning the cabin of two Cornishmen, situated on a high ridge on the treacherous slopes of Mount Davidson, was overwhelmed by a snowslide and swept a distance of fifty feet. James Nettle from Carn Brea, near Redruth, was lucky to escape naked from the mountain of snow, but James Northey was nowhere to be found. Digging frantically with his hands, Nettle came upon him in nine feet of snow; he had been buried alive, feet up and head down. The inquest revealed that he had no

---

1 Lord, p. 183.
2 Lord, p. 231.
3 Smith, p. 59.
4 V.C.T.E., 26 March 1867.

relatives and few possessions—only eighteen days' wages due from the Yellow Jacket, a watch, a trunk in which he kept his best clothes, and a few old letters[1].

Time and time again the story is one of marginal living, even in the boom years, and of the careful hoarding of a few precious possessions against petty thieves who ransack cabins while the miners are below ground. Twice the Cousin Jack Treweek returned to his shack to find his little world gone: "six bits, a carpet sack of old letters, wildcat stock and an old watch"[2]. Some, it seems, slept rough and died paupers: "John Thomas, a destitute Cornishman, sent to the County Hospital two weeks since by the City Relief Committee, died last Saturday"[3]. And often those in work provided the only relief for the dependants of the miner killed. Take William Truscott, forty years of age and pump-master at the Crown Point mine, "a steady temperate man, never losing a day" who "was in the regular habit of sending money to his wife and children" in Cornwall. One fatal day he was reaching for his tools when he was hit by an ascending cage and thrown into the main shaft. A heavy man, weighing thirteen stones, he broke through a platform of 3 in timbers, fell 400 ft and was picked up "a perfect jelly of broken bones and bruised flesh". The inquest revealed that his family was "in a poor position financially", so the Cousin Jacks at Gold Hill subscribed their dollars and sent them off to Cornwall.[4]

Where Death was the great leveller in the blind underworld of the mine, churches and playhouses staged again this drama of men struggling against the relentless forces of Nature. By 1862 Catholic, Protestant and Methodist churches were in position, ready to serve the living and to administer the last rites to the dying. Within shouting distance at Piper's Opera House, the finest actors played the lives and deaths of the princes of another age and time, Hamlet, Othello and Richard the Third[5]. The Prince of Denmark invariably filled the house, especially when the lead was Edwin Booth, for drama was the stuff of their daily lives. It was hardly different from a fight in Uhl's saloon or a duel between Cornish and English wrestlers arranged by Joseph Williams of Grass Valley[6]. And on one occasion Booth wanted a "practicable" grave for his last act and engaged two Cousin Jacks to do some pick work through a section of the stage. So professional were the results

[1] Ibid., 23 February 1867.
[2] Ibid., 26 March 1867.
[3] Ibid., 18 April 1867.
[4] Ibid., 17 July 1867.
[5] Ibid., 11 May 1867.
[6] Ibid., 31 May 1867.

that, when the gravediggers struck the skull of poor Yorick, they shovelled real ore on to the stage; and when Hamlet and Laertes jumped into Ophelia's grave, they barked their shins on real bedrock[1].

But, like the actors and the lecturers, the Cousin Jacks came and departed, moving along the lodes from the Savage to the Chollar-Potosi, from the Yellow Jacket to Gould and Curry, or disappearing altogether without trace. Nowhere else in America did population change so rapidly as on the Comstock. This was due to the frequent runs of bad luck, the severity of the winters, the ease with which wages were lost in the sixty-nine saloons of Virginia City and the twenty-one in Gold Hill, the demands of the four physicians who patched up the lucky, and of the undertakers who buried the unlucky and gave special attention to the grisly business of disinterring bodies and preparing them for shipment. But by 1868 the most unpalatable fact was that the best of the bonanza had been harvested and the rest simply dumped and forgotten or even used for paving roads. About 45,000,000 dollars' worth of silver had been taken down to the banks in San Francisco, but the gross and inept wastage of the rest played havoc with men's lives. The tapping of new veins declined every month; the value of property declined at the same speed; and the companies stopped paying dividends. So, as the slump settled, Virginia City once again looked like a stricken Sodom and Gomorrah. The miners who remained were family men. After all, children had to be educated, and Virginia City was rightly proud of its thirty-nine schools and its forty-eight teachers who cared for the differing needs of nearly 2,000 children at a cost of 32.50 dollars a head per year[2].

For many of them who braved the terrible winter of 1868, sheer destitution was only just round the corner. There was a scarcity of food, prices were high, ready money was short. There was hardly any wood for firing, the Chinese peddling it at the rate of 1.50 dollars a donkey load, just enough for six day's burning in a medium parlour stove[3]. Their wilderness of snow and ice was as implacable as the mines they had been forced to leave. Mark Vial, "native of Cornwall" and twenty-five years old, was fished out dead from the Carson River near Island Mill[4]; and Richard Copeland was found stretched out stiff beneath a mound of snow[5]. John Williams, "native of St. Just parish, Cornwall, 26 years, 5 ft 10 ins, dark complexion, spare built, goatee beard",

1 Drury, p. 54.
2 V.C.T.E., 17 January 1868.
3 Ibid., 15 January 1868.
4 Ibid., 28 January 1868.
5 Ibid., 16 January 1868.

simply disappeared after leaving the Occidental mine to collect his washing from a Chinese laundry[1]. Zachariah Curnow pondered over other problems, perhaps not unrelated to the strains that the winter produced in the women: "Whereas my wife, Mary A. Curnow, having left my bed and board without just cause or provocation, I will not pay any debts"[2].

The supply of labour still far exceeded the demand, nevertheless we find some of the Cornish courageously cutting their last threads with home and becoming American citizens: T. Polkinghorn, John Goldsworthy, James Hocking, William Mitchell, John Bishop, James Opie, James Pascoe, John Gribble, John Penberthy and James Tresize[3]. Miners were becoming more expendable than timber, for they were more easily replaceable, as the Opie cousins discovered to their cost. Peter Opie, at work in the Occidental, complained of the poor state of the timbers. The next night they collapsed and killed his cousin Thomas Opie. Philip Richards, who was working alongside him, said at the inquest, "I heard the ground fall, and I looked round and saw that his light was out"[4]. At the sinister Yellow Jacket thirty-four miners were burned to death in a single catastrophe that shocked the entire Comstock. Rescuers, "groping their way with dimly burning lanterns through the utter darkness and choking atmosphere of the drifts", came upon a gruesome sight: "dead men were lying on the floor of the level as they fell in the agony of suffocation with their mouths glued to cracks in the planks or raised over winches, turning everywhere for one last breath of fresh air"[5]. It was in this same mine that Joseph Oates was working by a chute when a large stone fell seventy feet and embedded itself in his brain[6]; and where Richard Stephen, in a moment of carelessness at the 800 ft level, simply forgot to place a plank across the mouth of the shaft, reached over it for his lighted candle and fell 300 ft to his death[7]. In fact, the accident rate at the Yellow Jacket soared so high that it was seriously suggested that either the management would have to pay danger money or draft men to the stopes[8].

As the slump continued on its headlong course, the fight for survival intensified[9]. The Hope mine at Silver City went bankrupt with liabilities

1   Ibid., 21 January 1868.
2   Ibid., 17 January 1868.
3   Ibid., 12, 21 April 1868.
4   Ibid., 10 July 1868.
5   Lord, p. 273.
6   V.C.T.E., 1 April 1871.
7   Ibid., 1 August 1871.
8   Ibid., 10 September 1868.
9   Ibid., 26 August 1868.

of 80,000 dollars and could not meet the wages of its 120 miners[1]. More and more Cornish appear in the lists of tax delinquents: S. Blight, Henry Peters, John Rosevere and William Tremain. A wave of crime swept over Virginia City and through the gulches, blazing houses covered up the tracks of burglars, and there was talk of a revival of summary executions by vigilantes[2]. Life above and below ground was more than flesh and blood could stand, the extremes of heat and cold striking down the most robust miners with acute pneumonia. By heavy doses of quinine, as much as 120 grains every twenty-four hours, and the free use of stimulants, the worst attacks might be repulsed; but all too often they ended fatally since the miners drank so much alcohol that its value as a stimulant ceased to be effective[3].

Fortunately, after five years of misery, the worst was over and a boom was in sight so electric that it would make Virginia's short history seem like the power of one candle flame. During the depression two mining engineers, MacKay and Fair, pooling their experience and the dollars of two saloon proprietors in San Francisco, had been testing their theory that the veins they had been tapping so long would eventually lead them to the core of the bonanza, deep in the heart of Mount Davidson. In March 1873 they found it at a depth of 1,167 ft and this time it really could be called "the Big Bonanza", for the vein was a gigantic jugular, 400 ft deep and from 150 to 320 ft wide. Virginia City was crazed with excitement as the rush began for the stock of the two dominating companies, Consolidated Virginia and Consolidated California. In five years their returns were staggering, over 100,000,000 dollars' worth of silver and, an accumulated stock of almost 160,000,000 dollars. For three years they were actually paying out monthly dividends of over 1,000,000 dollars! So everyone again owned stock and played the old investor's game of "you tickle me and I'll tickle you". Even the Cornish found that some of this wealth had rubbed off on their sleeves, one housewife complaining to her neighbours that her husband had vanished to Cornwall with their joint winnings of 4,500 dollars[4].

By now, according to John F. Uhlhorn's *Virginia and Truckee Railroad Directory* of 1873-4, there were about 100 Cousin Jacks and their families in Virginia City and another 130 at Gold Hill, about two per cent of the 30,000 reputed population. Employed at the Chollar Potosi, the Gould and Curry, the Belcher, the Ophir, Hale and Nor-

---

1   Ibid., 10 February 1871.
2   Ibid., 15 March 1871.
3   Lord, p. 374.
4   V.C.T.E., 10 March 1871.

cross, the Savage, the Crown Point, the Sierra Nevada, the Consolidated Virginia, the Julia and the notorious Belcher, they provided many of the foremen, captains and superintendents, even though they were outnumbered by the Americans, Irish and Germans. When Eliot Lord visited the Comstock in the early 1880's, he found them to be "all picked men, young, strong and vigorous, fed on the choicest food which the Pacific coast supplies, and paid the highest wages earned by any miner in the world". Choice cattle, "fatted on the succulent grasses of the Truckee meadows", were slaughtered for their tables. Fresh vegetables from the valley of the Carson were brought daily to the mines, venison from the Sierra foothills, wild-fowl from the California estuaries, fish from the Pacific and fruit of every description in abundance. Few, he says, spent their money on costly jewelry, but wore "shapely clothes", which they discarded before they became threadbare. Their cabins and cottages were unpretentious but well-furnished with bright coloured curtains, soft carpets and "prettily figured wallpaper"; and the water-closets were near at hand. Their churches were well attended, the schools excellently maintained and the teachers highly paid; in 1880 only four per cent of the children failed to attend[1]. Their library at the headquarters of the Miners' Union contained 2,000 volumes; and we know from other evidence that it was well patronised by the Cornish[2]. But in the early days of the new boom the cost of living soared astronomically. Some invested their savings with one Dunlap who absconded with his mistress[3]. Thieves were ever prowling the cabins of absent miners: four Cousin Jacks, who kept "bachelor's hall" on North B Street, found their cabin ransacked and 178 dollars missing[4]. Even six months after the Big Bonanza came in, stocks were still falling, due, it was believed, to the weakness of the Ophir mine; and Cornishmen felt the pinch as the Belcher, Crown Point and the Yellow Jacket paid off the men[5].

Though they had not yet reached the class of Roman gladiators (Lord's term of approval), yet the Cornish staged spectacles that almost put them in that category, pugilistic festivals in which they matched themselves against Celt or Saxon, provided the stakes were worthwhile. By far the most renowned of the Cornish boxers was James Trevillian from Camborne. He was twenty-eight when he appeared at Virginia in 1871. He had worked beside his father at the Forest and Dane mine

[1] Lord, pp. 313, 368, 372, 373.
[2] Drury, pp. 70-1.
[3] *The Virginia City Evening Chronicle,* 29 March 1873.
[4] Ibid., 7 April 1873.
[5] Ibid., 5, 7 March 1874.

in the iron regions near the source of the Hudson River, seen him blown to fragments in a nitro-glycerine explosion, and tested his own muscles in the gulches of Colorado. His reputation with his fists so impressed the Cornish promoter Bing Williams of Grass Valley that he offered to train him at the Sutro Saloon, kept by a retired Cousin Jack, Johnny Rowe. His two most memorable fights were against the Irishmen, Patsy Hogan and a man named Sullivan. He had Hogan "bested" in the seventh round, when a spectator shouted that Trevillian was wearing a piece of iron in his glove. The referee stopped the fight and found in his glove, not iron, but a piece of oakum, two inches long, that his seconds had placed there because his hand was swollen. The referee had no alternative but to disqualify Trevillian, whereupon all hell broke loose. Pistols were drawn and shots fired; but, in the phrase of the day, he "failed to get killed". The encounter with Sullivan was beset with difficulties. There was a ban on prizefighting and the sheriffs of Ormsby and Lyon counties threatened to arrest both promoters and contestants if the fight was staged in the areas of their authority. So a special train, hired from the Virginia and Truckee Railroad Company, left Virginia at eight o'clock one March morning in 1876 for an undisclosed destination, anyone being allowed to board it who possessed five dollars. Outside Gold Hill, however, the two sheriffs galloped alongside, so "all steam was crowded on and the train shot by like a flash, leaving the officers behind". At 9.30 the train hissed to a halt outside the Empire mine, and a ring was hurriedly assembled, but to no purpose. Sheriff Swift of Ormsby County, living up to his name, raced up just as the contestants were about to begin. Undaunted but ill-tempered, the miners pulled up the ring, clambered on the train and ordered the engineer to drive into the desert, where no sheriff would dare to interfere. But they were mistaken, for Sheriff Cook now appeared, just after they had completed their previous procedures. This time they refused to disband, hoisted him on his horse, watched him ride sadly way and then spent an hour arguing who should referee. They need not have bothered, for the contest was over in seconds, Trevillian's first blow landing under Sullivan's left ear and knocking him senseless. They collected twenty-five dollars for the Irishman and lifted him on the train. But he was never the same man again, and neither was Trevillian. A year later this "victor of the prize ring" was drinking in Brandt's saloon in Eureka when he started shaking as if he had the ague and fell down dead. He was mourned throughout Nevada because "he was a man of good impulses and made many friends"[1].

1 Ibid., 4 January, 6 May 1873; and 21 June 1877.

Men drugged themselves with these spectacles to help them forget the "rock furnaces" below ground that broke their bodies and stretched their nerves to the point of snapping. They wore only a light breech-cloth around their hips and thick-soled shoes to protect them against the steaming water that trickled over the floor of the levels. They worked in an intense heat that was made worse by decaying vegetable matter, hot foul water and the stench of human excretion. Tons of ice were sent down daily to the stopes, where half-fainting miners would chew fragments to cool their parched throats; on an eight-hour shift one miner would drink three gallons of iced water and consume 95 lbs of ice. It is no wonder that they staggered, bent double with agonizing cramps, raving and shouting incoherently until they had to be carried to a comparatively cooler part of the mine and the pit of the stomach rubbed vigorously. Such was the sacrifice demanded by the silver goddess; and the deeper the mine, the more she exacted from her votaries. Four men would be required to perform the work of one, and time was always running out as timbers threatened to collapse under the weight of the crumbling and swelling masses of feldspar, forced out of position by the shifting ground and rotted by the intense heat and moisture. At the Yellow Jacket in 1879 work had to be stopped because the thermometer reached the unprecedented level of 134°F, nine degrees higher than the previous record, that at Wheal Clifford in Cornwall. When the miner reached the surface, after travelling in the swift cage a distance of 2,000 ft in about three minutes, as likely as not he would find himself in a snowstorm. Such a rapid change of temperature and atmosphere would bring on waves of dizziness and nausea and, if friends did not support him, he would collapse against the timbers of the shaft and fall backwards into the waters of the sump[1].

Physical weariness, nervous exhaustion and economic anxiety all produced their strains and stresses and often made men unpredictable. Shortly before Christmas 1874 four Cousin Jacks sat round a table at Washington House to play their usual game of pedro: Thomas Trembath, Joseph Hodges, Michael Roach and John Skewers. About midnight they were joined by Alfred Rule from Camborne who suggested they should play for drinks instead of "for pastime", a mistake as it happened, for it made them quarrelsome. When only Trembath, Skewers and Rule were left in the game, Rule suddenly accused Skewers of cheating by looking at Trembath's cards. In a flash tempers

[1] Lord, pp. 321, 389, 394, 396, 399.

flared, Skewers denied the charge and Rule called him "a goddammed liar". Rule, being heavily built, struck Skewers across the shoulder with his left hand, grasped him by the collar of his coat and forced him backwards on to a table top. Impulsively, Skewers dragged a pistol from his pocket, fired it at the back of Rule's left ear and killed him instantly. The shocked onlookers were the more horrified since Skewers was usually a mild man who had never been known to quarrel. It was in keeping with his character that he at once surrendered to the sheriff, while sympathetic Cousin Jacks somehow raised no less than 10,000 dollars bail[1]. He was tried for murder in the spring of 1875, but the jury failed to agree, so a fresh trial was ordered and the bail reduced to 6,000 dollars. Unfortunately a recession had set in, the population of Washoe had dropped from 1,500 to 500, and the money could not be raised, so into the county jail went the unfortunate Skewers to await trial in the following September[2]. For some reason or other there were further delays and then, in the last days of October, began the terrible fire in Crazy Kate's lodging house on A Street that wiped out almost the whole of Virginia City. Skewers now disappears completely from the pages of the newspapers, except for a note that, at the height of the conflagration, he was lodged for safety in a tunnel of the Sierra Nevada mine and then in Gold Hill jail. As for poor Rule, he was almost forgotten, for his friends were so terrified that none dared tell his relatives in Camborne of his fate for a year.

Virginia City had died once before, in August 1863, when a fire destroyed over 7,000,000 dollars' worth of property in the business section. Dreadful as that was, however, it could not match this second holocaust which, with the onset of winter, reduced over 2,000 homes, business houses, hoists and mills to charred rubble, and split the canyons with the explosion of ignited blasting powder. Though the plant of *The Virginia City Territorial Enterprise* was gutted, the staff on borrowed machines managed to produce on 28 October 1875 a single sheet that wept over a city in ruins where, "amid the universal wreck, people wandered through the débris of Virginia City yesterday with such a look on their faces as men and women wear when they gather round a coffin". Within twenty-four hours, stocks and shares fell by 25,000,000 dollars, women and children were sheltering in schoolhouses, legal cases had to be suspended, and so widespread was the misery, not least that from the pall of smoke and mercury fumes causing everyone to cough and spit, that free passes were issued on the

1   V.C.T.E., 22 December 1874; 10 April 1875.
2   Ibid., 7, 15, 20 May 1875.

railroad to those who could make their way to friends in California. For the survivors forced to remain the major problem was winter. On 14 November the Washoe "zephyr" howled down the mountain side with the power of a tornado, demolishing the charred remains and shells of old buildings and sweeping away the temporary shacks. Blinding snow-storms harried the thinly clad who were huddling for comfort in make-shift shelters. Rivers of foul mud oozed down the streets, rats darted from their holes, and typhoid did the rest. Railroad communications with the outside world were crippled; a freight train with supplies was derailed near Reno, and another, loaded with refugees, was halted by a snowslide near Clipper Gap. Unclaimed letters piled up in the Post Office for A. H. Barrett, R. Liddicoat, F. H. A. Mitchell, W. H. Odgers, W. Penrose, W. H. Pellow, Jonathan Pope, J. Trewarkus, William Skewers, Jonathan Teague and many other Cornish who were never seen again in Nevada. January brought no relief. A supply train, pulled by twelve engines, failed to force its path through the snow, as did the plough that was despatched to dig it out. An avalanche rolled down Silver Mountain and engulfed the cabin of three Cornishmen, partially destroying the Exchequer hoisting works on the way. John Mitchell freed himself after "terrible efforts" and it took two hours to release William Bartle who, though alive, had lost his reason. John Craze, William Angove and S. Tregloan could not meet their tax demands; and Amelia Trembath divorced her husband because of his "failure to provide the common necessaries of life" for their children[1].

For the next five years the tale of the Cornish Comstock unfolds a mood of increasing melancholy as ordinary people tried to live a normal life under abnormal conditions. John Daley's end was perhaps exceptional and he hardly deserved much sympathy, but that of Mrs. Wicks was a bitter comment on the anxieties of women as they tried to adjust themselves to the wilderness of the mining town, the ravages of Nature and the dreaded daily procession down C Street of black plumed horses and draped hearses.

Daley "from Cornwall" ran a saloon called the "Bon Ton" in which he was rarely seen without a pistol and bowie knife in his belt, a comment no doubt on his quarrelsome nature, though he was usually "industrious and steady". But he made no secret that his sworn foe was one R. H. Carter because of their mutual attachment to the same "public woman", as Katy Twist described herself. When she went to

[1]   Ibid., 12, 28 January; 15, 25 February 1876.

live "privately" with Carter, which may have denoted some return to respectability, Daley was incensed. One Sunday night found him in a particularly vengeful mood but vaccilating between the influences of saloon and chapel. Unfortunately the bottle triumphed over the hymn-book and, towards midnight, in the company of Stephen Richard who had served a jail sentence in Carson State Prison for "shooting a fellow", he presented himself at the bedroom window of Carter and Twist and threatened he would "get his son of a bitch". A gun was drawn, three shots were fired, and Daley fell to the sidewalk with one of them lodged in his heart. Needless to say, the jury brought in a verdict of justifiable homicide and Carter lived to take Miss Twist to bed on many another night[1].

More to be pitied than condemned, Mrs. Wicks was a victim of the trying conditions that beset the mining industry in 1877 when the Virginia City newspapers were reporting: "The streets are thronged with miners out of employment; hundreds of families are absolutely starving on the Comstock, and the prospect grows more desperate daily"[2]. She had several children, her husband and her brother were both out of work, and she was pregnant again. Out of sheer desperation, while her husband was trying to get a job away in Grass Valley, she attempted an abortion. Reports suggested that she was not alone in her predicament and that a professional abortionist was at work in the Union saloon, kept by a prostitute called Carrie McCormick. The "woman doctor", whom Mrs. Wicks consulted, seemed hardly better, for under interrogation, she claimed that she had been practising medicine for fifteen years and had passed her examinations at the age of sixteen in Santa Monica! According to the evidence of two of her Cornish friends, Thomas Ninnis and Mrs. Penrose, Mrs. Wicks permitted herself to be immersed in a bath of hot water and an electric battery applied. As a consequence she died almost instantly, a tragic example of an attempt to defeat the hunger that gnawed constantly at the stomachs of whole families in a city of wealth where the accident rate was so high in 1877 that it provided employment for thirty-five physicians and surgeons, about one to every 300 souls[3].

Cages were raised and lowered so fast that a moment of carelessness could result in a severed leg, arm and head, or a bleeding body that was no more than a mangled mess of flesh and bone at the bottom of the shaft. John Sinnott was crushed so violently between the cage and

1   Ibid., 25, 26 April 1876.
2   Ibid., 2 May 1877.
3   Ibid., 3, 11 March 1877.

the floor of the Savage that his ankle bone was driven clean through his indiarubber boot[1]. Edward Champion, a respected member of the Pachahontas Tribe of the Improved Order of Red Men, fell from a moving cage and was killed instantly[2]. William Lidicote from Crowan broke his neck down a shaft at the Ophir[3]. William Eddy's job in the Combination shaft was to watch the water level in a tank; it was neither dangerous nor exacting, but so monotonous that, in the hot humid atmosphere, the temptation was to drowse. One night on the graveyard shift he woke up suddenly, lost all sense of direction momentarily and walked into the shaft, falling 325 ft into three feet of water. Though he was killed instantly, his relatives considered themselves fortunate for he was brought to the surface in one piece and strangely recognizable. Miraculously his breathless body had plummetted to its destruction without touching the sides of the shaft, his rubber overcoat acted as a parachute, and the water braked the last few seconds of his fall[4]. Not all were so favoured by the gods. Usually only the undertakers were permitted to view the remains who hurried to stake their claims on the morbid principle of "first come, first served".

Such was the fate of young John Evans, who fell from the cage in the Sierra Nevada and was returned in the same cage in a blanket, a horrifying bundle of unconnected sinew and bone. They buried him at the same time as Charles Delaney, the President of the Miners' Union, and the same impressive procession did for both. Masons attended resplendent in uniform and regalia, miners and bartenders followed in their Sunday black, and the Knights of Pythias were headed by the band of the Washington Guard with their drums muffled[5]. Nowhere else did the canyons roll so often with the thunder and thud of the marches to the dead heroes of Cornwall, and many a bare-headed bystander wondered when the bells would be tolling for him. John Morrish's head was crushed to pulp when he made a false step on the Chollar Potosi cage[6]. John Williams was picked up so shockingly broken that not a single bone was in place; Richard Noall, at "croust time" at three o'clock in the morning, fell into the sump containing nine feet of scalding water; both of them left families to mourn them in Tuckingmill and St. Blazey[7]. Even more terrible was the end of William

1   Lord, p. 402.
2   V.C.E.C., 24 February 1873.
3   *Gold Hill News*, 29 November 1865.
4   M. M. Matthews, *Ten Years in Nevada or Life on the Pacific Coast* (Buffalo: 1880), p. 314.
5   V.C.T.E., 17, 22 September 1874.
6   Ibid., 1 November 1874.   7   Ibid., 29 January and 18 June 1875.

Jenkins who fell into the sump of the Julia mine in February 1879. The temperature of the water was 158°F and, though he was immersed for only a few seconds, he was flayed alive, "pitifully crying for death, while his anguished friends tried to relieve his sufferings with morphia"[1].

The mine harboured respect for neither the young nor the old, for the experienced nor the novice, when a momentary lapse of caution was more than sufficient to hurl them to perdition. One accident which sent "a thrill of horror through the community" occurred when the signal rope got caught in a loose timber on the side of one of the shafts in the Consolidated Virginia. John Trembath was riding in the tub at the time. The rope ripped out the loose timber, which fell 150 ft, hit Trembath full in the chest, tore open his right side and passed like a spear through his liver. Thomas Cook from Bodmin, "one of the oldest and most experienced miners on the Comstock", fell 700 ft down the main shaft of the Knickerbock. Joseph Pierce broke his neck when standing on a plank that gave way beneath him[2]. At the Savage, John Moyle and Matthew Eddy were riding the cage at the very moment when the engineer, J. F. Mitchell, came on duty. Making the appalling error of thinking that the cage was just beginning its ascent, he put on full steam; it crashed into the sheave with a fearful sound and pulverised Moyle beyond recognition. He was thirty-five, with a wife and five children in Cornwall dependent on him[3].

The Washoe mines were so deep and the ground shifting so constantly that death by cave-ins or from collapsing timbers was almost an every-day occurrence; William Barnett in the Crown Point, Alfred Cox in the Sutro Tunnel and Henry Goldsworthy from Newlyn were all dug out dead from runs of waste rock and rubble[4]. But some of the most frightful accidents were caused by the "giraffe", a long platform car with low front wheels and high rear ones, which ran up and down a tracked incline carrying the ore, its movements governed by a cable. In the Yellow Jacket, the cable suddenly snapped, causing the platform to run back with sickening velocity into a group of miners. Four Cousin Jacks, Charles Bennet, Samuel and Thomas Odgers and Francis Polkinghorn, were severely injured, but Richard Pearce had the back of his head torn completely off, while "the face remained a mangled mass which looked like a mask". Only twenty-two, he was one of their most promising lay preachers in the chapel down at Gold Hill[5].

1 Lord, p. 399.
2 V.C.T.E., 20 July and 8 November 1875; 19 March 1876.
3 V.C.E.C., 4 September 1877.
4 V.C.T.E., 10 May 1876; 7 March 1877.
5 Ibid., 3 May 1874.

If mining accidents on the Comstock can be classified and graded into degrees of horror, then the most frightful was fire. It was in one subterranean hell that several miners were building a bulkhead to prevent a fire from reaching the stopes. Suddenly the ceiling caved in. In an instant all the heat and flame in the air-shaft were sucked downwards and sideways into their drift. The deluge of fire lasted only a moment, for almost immediately it was sucked back into the shaft. In that flash they were scorched to death in an agony of pain, their cotton overalls reduced to ashes and the skin peeling off their faces and bodies in great sheets. And among them was the writhing body of the young Cornishman, William Johns, whose shrieks of pain pierced the mine for several hours before he mercifully died. The cause of the disaster was found to be simple: a candle left burning near explosives.

Against this sombre background something more was needed than nerve and the saloon to meet the perils of the next shift, especially if the miner had escaped disaster only by a miracle. Jacob Laity, shift boss at the Belcher, came out alive from a hurricane of fire along a floor littered with the dead bodies of rats[1]. Jack Bluett was dug out alive from a cave-in that crushed three sets of timbers in the Sutro Tunnel[2]. Joe Trewhella was hit by a blast at the 800 ft level of the Justice mine and survived[3]. Only a grim sense of humour could steady the nerves after near-calamities like these. Elijah Billings was at work in the main shaft of the Imperial when he fell into the pump compartment. His companions were sure he had pitched 1330 ft to certain death, but by a miracle he had managed to cling to the pump bob where it joins the rod. When he had been pulled to safety, his only observation was: "By the bloody 'ell, old son you, if I hadn't caught 'old of the bob, I'd been scattered all abroad"[4].

Accidents seemed to come in waves and at random, with no apparent connection between them. Hardly a week passed but the undertakers were busy in their parlours straightening out some hard-rock stiff who now knew that his hole was deep enough. Now and again the newspapers demanded that the companies should introduce some pension scheme or that the miners should be encouraged to be thrifty. But in most cases it was the charity of friends that averted destitution, every miner "contributing liberally as he draws his wages"[5]. Little could be saved out of 4 dollars a day, for the cost of living was high, though

---

1   V.C.E.C., 20 September 1873.
2   V.C.T.E., 14 April 1876.
3   V.C.E.C., 29 September 1877.
4   V.C.T.E., 31 October 1876.
5   Ibid., 3 September 1876.

some Cousin Jacks were earning much more: chief engineers 10 dollars; boss carpenters 8 dollars; blacksmiths 6 dollars. But only a few reached the very top income bracket where superintendents and mill managers enjoyed such high salaries that they looked almost like medieval princes[1]. One observer thought the Cornish feckless for, though they were more hardworking than the Germans, the Americans and the Irish, they were the poorest, for "drinking and gambling are the cause of two-thirds of the suffering in Virginia City"[2]. Despair inevitably led to the saloon when men saw their companions dying so far away from their families. William Pope never lives to see his wife and six children again in Cornwall, dying of heart failure at the age of forty-seven[3]. Richard Gluyas digs out three Cousin Jacks from their cabin that has been wrecked by a snowslide; Thomas Champion is alive, but Isaac Jewell and Moses Willey are both dead. They come from Helston, and, in some cottage there, Willey's wife and seven children await the last of the monthly remittances[4]. And often to the survivors fell the unpleasant task of acquainting relatives about the disaster long after it had happened.

By 1877 the drift from the Comstock had begun, just as *The Territorial Enterprise* had predicted two years earlier: "It is a sad day when the American citizen can no longer go west, but we have seen it coming for a long time"[5]. Yet still they hopefully trundled in from the east, some to test their strength and convictions in the Black Hills[6] and others who would soon hurry on to California, where they would hide their disappointment on finding little work in the mines by farming precariously for as little as 20 dollars a month. To these greenhorns the Comstock still looked attractive, for the monthly silver yield in 1876 had been 7,000,000 dollars. But before the year was out single men were being sacked right and left and married men only retained as long as funds permitted[7]. Some of the bachelors struck out for Panamint City on the edge of Death Valley, where conditions proved to be about as rough as in Nevada; it was no more than a make-shift camp of twenty-six frame buildings, stockades and tents with a couple of murders already on its hands. Some of the married men wasted no time in sampling the much advertised New Almaden quicksilver mine

1 John J. Powell, *Nevada the Land of Silver* (San Francisco: 1876), p. 276.
2 Matthews, pp. 169, 172, 173.
3 V.C.E.C., 24 July 1873.
4 V.C.T.E., 19 January 1875.
5 Ibid., 9 April 1875.
6 Ibid., 11 March 1876.
7 V.C.E.C., 7 July 1877.

and the lush Santa Clara valley, where J. B. Randol needed them, and Grass Valley. Prospects were still good here, according to the Cornishman Polglase, foreman at the giant Idaho-Maryland mine, when he visited the Comstock to inspect the "ponderous machinery" of the Consolidated Virginia'[1]. Not a few made the mistake of setting their sights on the new silver camp at Bodie where by 1879, some fifty mines had started; within two years they were no more than holes in the sand surrounded by sagebrush[2], compelling John and Mary Mitchell to swing out again on the long orbit to security that had begun in Australia and was to finish in Nevada City. Some travelled hopefully to Delamar and its unsuspected snares. Though the wages here were high, it was a dry mine of hard quartzite which threw off so lethal a dust that the youngest of the Cousin Jacks soon succumbed to silicosis, and the town became known as "the maker of widows", where black dresses and black hats were the prevailing style, so frequent were the funerals. Some miners vanished and were never seen again, like the uncle of the nine-year old Richard Jose. And some perhaps survived the depression by raising beef cattle in the lush meadowlands of Nevada.

Today the Cousin Jacks of the Comstock will only be remembered perhaps in their eleven cemeteries on Decoration Day at the going down of the sun over the vast empire of the relentless desert; or at the museum at Virginia City, where you can see a working model of a Cornish pump and the beam of another that was installed in the Kossuth mine at Silver City, whose pump rod of massive timbers regularly and evenly stroked its path through the blackness of all the 900 ft of its length. Yet the Cornish have not quite disappeared from the face of the Comstock. While some Americans hive to Nevada to gamble and others to savour once again the endless transformation scene of the frontier, the Ninnises from far-off Gunnislake have anchored themselves emotionally and economically to the bare soil around the Silver City that was. All their relatives are buried here in the canyons that were once alive with the shouts of brave but poor prospectors. They remember the Emma Nevada mine that never paid although it yielded pockets of white quartz held together by "wires of gold". They recall their attempts at leasing and the long spells between "clean-ups" when there would be no income. The Church was their strength and stay and the daily reading from the family Bible their guide to right action; and when the Church was burned down, their frame house on Main Street took its place and

---

[1] V.C.T.E., 11 October 1877.
[2] Smith, p. 217.

The view from Mount Davidson of Six-Mile Canyon and Virginia City, Nevada, about 1900. [Nevada State Highway Department]

Silver City, a Cornish town in Nevada, as it looked in 1890. The dirt road leads northwards to Mount Davidson and Virginia City. A more desolate region it is difficult to imagine for emigrants bringing up a family.
[*Bancroft Library, University of California*]

Part of the industrial complex of Virginia City: the Consolidated Virginia's pan mill, battery and hoisting works, at the height of the Comstock silver bonanza.
[*Bancroft Library, University of California*]

their living room table became the lectern and pulpit for the preacher from Reno or Carson City. Frederick Charles Ninnis and his wife, now well beyond the age of normal retirement, are still hardy prospectors. They are conscious of their Cornish background and the struggles of the pioneers who died young, either in the mines or from the pure white brittle quartz from the Yellow Jacket that cast its deadly clouds of powdered crystals into the air and the lungs. Such was the heavy price that the Sun Mountain exacted from those who cut its veins and pierced its silver heart.

VII

FOR a few Cornishmen who had sweated their lives away among the rocky pinnacles of split granite and the parched deserts there was one part of America whose call to them was almost mystical: Oregon. Blessed with one of the world's most beautiful rivers, the Columbia, mountain peaks of matchless beauty like Hood and the Three Sisters, the crystalline depths of Glacier Lake and Crater Lake, rich farmlands of cattle and fruit, extensive rain forests of fir, pine, hemlock, spruce and cedar and 400 miles of silvered sandy beaches, Oregon was "Sweet Home" (as one town has aptly called itself) for those who had travelled along the world's longest trail. Poised on the edge of the continent, it seemed a paradise where a miner might live the most rewarding and satisfying of lives, clearing ground and sowing crops by the margin of its well-stocked, teeming streams.

Oregon has never strictly been considered one of the mining states, in spite of its deposits of gold, silver, lead, copper, zinc, uranium and chromite, its rich iron fields near Oswego and its beds of coal on Coos Bay. In the early days it only lay across the path of the gold hunters; its first settlers in the 1850's abandoned their ploughs for the lure of California; and disappointed prospectors of the Mother Lode gave it no more than a passing glance as they swept northwards to the diggings in Caribou in British Columbia. Apart from the freak placers near Coos Bay that were being worked in 1875, and the partially uncovered gold-quartz ledges in the southern areas of the Cascades, the discoveries of 1851 in Jackson and Josephine counties were really an extension of the Californian veins. Those in Baker county to the east were a continuation of the Idaho lodes beyond the Snake River, the mineral deposits running to and fro with a complete disregard for the niceties of state boundaries. The gold stampedes to these areas soon exhausted themselves and flung their miners back into the wastes of Idaho. Though Idaho was the mining magnet for the dispossessed Cornish of Nevada and Colorado in the 1880's as well, Oregon displays links with Cornwall that must be considered first.

Its heartland is still, as it always has been, the Willamette Valley, a stretch of 140 miles between the coast and the Cascade Mountains that has an English look about it, for the climate is pleasantly temperate all the year round, producing an abundance of fruit of all kinds. All that the English exclaim about their own countryside is here with the addition of the logging trucks that bring the timber from the vast forests to be manufactured into paper, cardboard and plywood; the river bows and bends through succulent meadows and meanders as lazily as the Wye around Chepstow through high green billowing fields of grass that the short summer sun turns into gold. This valley is the cradle of the history of the Pacific North-West. It began rocking hardly more than a century ago when men of the calibre of McLoughlin, Astor, Bonneville and Wyeth built their forts and fur emporia; and its first child was Oregon City, the earliest incorporated town west of the Mississippi, settled in 1829 by the English Hudson's Bay Company to head off American emigrants and to provide a supply base for its trappers. In those days the Englishman knew every nook and cranny for, disguised as sportsman, journalist or convalescent, he was engaged in a desperate struggle for survival against the spies of Spain, Russia and America.

Since Oregon ended an extremely long supply line, either from Churchill or London, the Americans experienced no difficulty in settling there without challenge. The publication of the diaries of Lewis and Clark, Washington Irving's account of the Astorians, Fennimore Cooper's *The Fur Traders*, the formation of innumerable immigration societies, the banking crisis of 1837, a severe outbreak of ague in the valley of the Mississippi, the devout wish of some to win Oregon for the Lord or at least to rip it from the imperial British, all played their part in filling Oregon with Americans. In 1843 the first caravan of more than 100 wagons pounded in from the east, their occupants wasting no time in deciding whether they should live under the American or British flags even though some members of Congress thought the rock-bound and harbourless coast quite worthless. However, the strategic with-drawal of the British, fearing for the safety of their stores at Fort Vancouver and admitting the decline of the fur trade south of the Columbia, removed all doubts. In 1846 the boundary with Canada was fixed at the 49th parallel and in 1848 the bill that created Oregon Territory was finally passed. Three years later, among the streams of immigrants that poured in, came a Cornish family from Wisconsin, bound for the rich meadows of the Willamette Valley.

The James family hailed from Trelan in the parish of St. Keverne, deep in the Lizard promontory which, for many travellers, is the first or last glimpse of England as they sail the English Channel. They were part of that remarkable migration of Cornish gentry, the Foxwells,

Moyles, Shephards and Lorys, who in 1842 made Wisconsin their destination. Samuel James, who had married Ann Maria, the daughter of William and Ann Harris Foxwell, followed them a year later with his wife and their four sons. Unlike the others in the tightly-knit Cornish community of Yorkville and Caledonia, Samuel James always seems to have been fired by visions of Oregon and the Pacific beyond; indeed his second son born in the log cabin he built on the banks of the Root River was baptized Richard Oregon. A voracious reader (he insisted on bringing from Cornwall his entire library, dead weight 800 lbs), he was no doubt influenced by the flood of popular literature about Oregon, and after seven years in Wisconsin decided to make the overland passage.

They left Yorkville in October 1850, many of their Cornish relatives warning James that it was dangerous and foolhardy so late in the year. But his intention was to spend the winter in Iowa, preparing for the main assault in the spring of 1851, which he considered would be a good year. He had made it his business to discover that emigration in 1850 had been too heavy for the country along the trail to support. The result had been unnecessary sickness and raids by the Indians. He calculated that in 1851 there would be fewer immigrants bound for Oregon, cattle feed would be more plentiful, and the Indians would not bother to interfere with small wagon trains.

So they crossed the Mississippi in a boat with "tread power and two horses for engine power", and came to Dudley, Iowa, a log cabin town of one hotel, saloon and store, where James rented a frame house for the winter and prepared himself for the venture across the continent: hiring teamsters; spinning and weaving cloth and then shaping it into garments; laying in sufficient supplies of smoked sausage meat; baking the main item of diet, "hard-tack", and packing it into 100 lb sacks. He bought three wagons, three yoke of oxen for each wagon and some loose cattle, but only three horses, for he had learned that oxen were better draft animals, seemed to gather strength as the journey lengthened, and required less time and care in their feeding than horses. His outfit, a mobile home that had to suffice for himself, his wife and their eight children, was part of a wagon-train that numbered twenty-six vehicles. On both sides of each wagon was painted a number from 1 to 26; the idea was that No. 1 would be in the lead the first day, and then would drop to the rear the next day to give each an equal chance of avoiding the dust kicked up by the animals. James also thoughtfully supplied himself with dark glasses, bound with leather round the rims, to reduce the sun's glare and to protect his eyes from the dust. To one

of his wagons he fixed a milometer of his own invention, his intention
being, not only to record the distance travelled, but to average about
twenty miles a day and so to rest on Sundays. Like many other Oregon
pioneers, James kept a log-book rather than a diary, his entries concerned with the practicalities of the expedition rather than with rhap-
sodies on the beauties of the countryside. He was only too well aware of
the dangers that lay ahead and the careful planning needed to ensure
that they would arrive at the right place at the right time[1].

The wagons started their accustomed rolling on 9 April and twelve
days later, over a "very bad muddy road all the way", arrived at Kains-
ville, near Council Bluffs, where a large contingent of Mormons had
already camped. Mrs. James did not like one of their groups, mainly of
women and led by an English "bishop", whom she roundly condemned
for being so depraved as "to mislead those poor ignorant proselytes";
but they were glad to buy from them their own guide-book of the trail
which contained the place-names that ultimately appear in the log-book
of James. After replenishing stores and buying buckskin clothing and
trinkets to trade with Indians, on 28 April they were away again under
the leadership of an old Indian scout, William Stone, crossed the
Missouri at Mormon Ferry and were soon being challenged by Pawnees
who demanded toll for passing through their land, riding alongside the
train in double file and concealing their bows and guns beneath their
buffalo robes. "Let them have some flour and two cows", reluctantly
noted James. Then for the next month, as they followed the Platte
River to Chimney Rock and Fort Laramie, they were harassed by
Sioux. They joined a Californian train for greater protection, made the
mistake of hurrying their cattle before they were properly fed and
watered, and then divided their wagons into two groups, so that the
Jameses now found themselves reduced to the dangerous level of ten.

Near Fort Laramie they pitched camp about a mile away in a village
of Sioux and Cheyennes and decided to rest for a week, "previous to
taking a long drive through an alkali country". So far their losses were
not alarming and there were some positive surprises for they had not
expected to come across a plains telegraph of buffalo skulls along the
trail on which immigrants ahead of them scrawled their messages.
James, ever resourceful, from broken bits of wagons built a two-
wheeled cart which, so he reckoned, was an improvement on the
heavier prairie schooner. Completing his preparations by trading some
of his hard-tack and sausage for bows, arrows and Indian ponies for

---

[1] The log book is in the possession of Mr. David James of Seattle, Washington. I
     am most grateful to him for giving me permission to quote from it.

the boys, and buffalo robes for all, James now entered upon the most difficult part of his enterprise west, for the Mormon guide-book had finished its work. From Laramie to the Dalles, the names of places and streams he mentions "are of our own coining".

This is not true perhaps of Willow Springs, where they came upon thick deposits of bicarbonate of potash that the women used for making bread, to give it a light and springy texture. Certainly it was not of the well-known Independence Rock, "as big as a city block two storeys high", on which James added his own name to those of emigrants who had passed this point of no return, his pen a brush and his ink the tar he used for greasing the wagon wheels. This country along the Sweet-water, where they could camp in the high sagebrush and their cattle could graze on the dry bunch or buffalo grass the other emigrants usually avoided, was reasonably well charted and populated; and James was even able to despatch a letter to Cornwall for fifty cents in the care of an Indian guide returning east. But the 120 miles to South Pass were soon to be transformed into such a wilderness of stone and grit that they would be forced to walk most of the way; and James sadly pondered on the thought that the Sweetwater was the last river whose waters would flow into the Mississippi, then into the Gulf of Mexico and finally be washed against the cliffs of Cornwall. After the Sweetwater all rivers would empty in the Pacific.

Bitter Cottonwood Creek proved to be all "alkali, alkali" where "bones of cattle whiten all the way", a country "full of poison". Between the Little Sandy and the Green River they encountered fifty miles of dry plain with no water, where it was so hot that movement was possible only at night. The cattle were so maddened by thirst that, when they smelled the Green River, they had to be unyoked and James feared they would kill themselves drinking, so swollen were their sides with water. Once over the deep and swift river (some Mormons exacting "ferriage" at the rate of 10 dollars a wagon), the oxen had to endure a long haul over "a dreadful rocky road" before they reached Fort Hall where, in a thick border of willows, James made his last purchases from the half-breed son of John Day before pushing out into the unknown. July was a harassing month; his cattle were stolen by Bannock Indians; the winding path of the Snake River provided nothing but desert and the only water was hot enough to scald. Currant Creek turned out to be "the most desolate country in the whole world, the region of the shadow of death".

Early in August they reached Burnt River, "a bad place to be attacked by Indians and much dreaded"; and it was here that James extricated his party from a menacing situation only by refusing to fire on an ever narrowing circle of Indians that spun around them. One family had not been so circumspect and were attending their father, shot in the bowels as he had gone to fetch water. But once over the Powder River and in "excellent grass", they found the Indians friendly and eager to trade their salmon for what was left of the hardtack, a surprisingly good bargain. But by now both oxen and drivers were feeling the strain as each new day took its toll of their strength: "got into Grand Round over a terrible hill, most of the drivers quaked in getting their wagons down". Others before them had faced defeat here, for the trail was strewn with wagon tires, iron fire-shovels, stoves, pots and kettles, and even an English roasting spit. But James's oxen were still expected to pull his library and, when the prickly pear of the cactus lamed them, he always seemed to have a spare half-shoe ready. Yet the way to the Dalles seemed melancholy beyond words; Mrs. James could not endure many new privations for she "always had servants in the old country and a hired girl or two in Wisconsin", the boys' cotton pants were torn beyond recognition and repair, the girls' hair was matted and tangled, and the clothing of all was beginning "to smell strong". But the most demoralizing experience of all was to come upon mounds of stones that were the graves of emigrants whose hearts had stopped from sheer exhaustion. Sometimes they saw a stake in the ground with an overcoat hanging on it; then they realized with a shock that a young man had only just died for this device was an investment against further molestation by Indians.

By the end of August their trail to Oregon virtually terminated with the flagstones of the Dalles on the Columbia. Whatever they did not need now until they reached Willamette was shipped by water to Portland, and for the first time James parted with his books at the small cost of 20 dollars. Courage and initiative however were still needed. Along Indian Creek they found themselves "in a deep valley and dreadful hills to go up and down". They slithered down "the muddy branch of des Chutes, a mile rough with rocks, roots and stumps", topped the crest of Mount Hood, and floundered in a "horrible mud-hole", which lost them a day in repairing the wagons. Finally on the sixth attempt they negotiated the difficult crossing of the Sandy River, climbed the summit of the Devil's Back Bone (where the wagons had to be let down by hand on ropes) and on 9 September 1851 entered Milwaukee, the small saw-mill town where sailing ships were being loaded with lumber for San Francisco. Here on the banks of the Willamette sprawled their Canaan, the land of the Big Trees, prosperous farms and friendly Indians, whom the Cornish at once dubbed the Amalekites. Appropriately their first night there ended with James reading the Twenty-Third Psalm. His milometer registered 1,890 miles since their departure from Council Bluffs five months earlier.

But the land he bought on the Clackamas River, delightfully sited though it was on a pleasant knoll, with a brook running to his door and within easy reach of Oregon City, proved disappointing. It lay in the uplands among the scattered fir timber, and clearing the trees by boring holes in their base and then setting fire to them was a heavy enough task. Even more backbreaking was uprooting the brush of high hazel and wild raspberry that was spread for twenty miles around. But not until the following Spring was it possible to know for certain that the land was useless, for it then became covered with, not the expected grass, but wild iris that the cattle would not eat. The truth dawned on James that he had arrived much too late for the best locations so, without hesitation, he decided to move on.

In September 1852 the wagons were once again loaded and driven to Milwaukee, the Willamette crossed at Oregon City and final plans made at Portland for a passage north. Here they were lucky to meet the great McLoughlin, who gave them a disused Hudson's Bay bateau. This they repaired, transferred all their baggage on board, and hired five Indians for 50 dollars to navigate them up the swift currents of the Cowlitz River. At Cowlitz Landing they stayed a couple of weeks with a John Jackson, "a warm-hearted Englishman known to every pioneer", awaiting the arrival of Samuel and a son. These two had driven the cattle over the trail and over the Cowlitz and Lewis rivers, a somewhat dangerous undertaking and, when they arrived at Jackson's, their clothes were in tatters and Samuel "had an old boot tied on one foot and a ragged shoe on the other". The journey northwards to the prairie village of Olympia, where he had selected a site of 340 acres "which the U.S. Government gave to all her citizens coming into the country after 1851", seems to have been equally difficult, for at one time they were reduced to eating the peelings from potatoes. It was the wildest part of country that had rarely seen a human being, so "this was the first time that a great fear came over Mrs. James when she realised that they had brought the family into a wild country where even the wild animals were not afraid of us".

But the new claim was ideal. It stretched for a mile along the banks of the Chehalis and its streams teeming with salmon. All around them the prairie looked like a fine lawn, skirted as it was by hills of fir and cedar, oaks and maple; and beyond rose the snow-capped Mount Helena. It was, however, already late October, and swift preparations had to be made for a winter they knew nothing about. Cedar logs were fashioned into bedsteads, tables and stools, and a door which was hung in position by leather hinges cut from Samuel's old boot. And from the same logs that winter they shaped a coffin for one of the boys who suddenly died of "a putrid sore throat" and, with a few Indians, watched the burying with muted horror. The first falls of snow lay unusually deep; a plague of crows helped themselves to the first sowing of wheat; and wolves attacked the calves. Yet the education of the children was never neglected, as one of them recounts:

I must tell you how we lighted our cabins at night; Abraham Lincoln had pine knots, but ours were fir, full of pitch. By this light all who were old enough studied their lessons, for father and mother were fine teachers, and no evenings were allowed to go without lessons; Latin and Greek and higher Mathematics were mixed with plain English.

The search for the bare essentials of a formal education had always been as difficult in Cornwall as in Oregon. Samuel James had chanced his luck 6,000 miles from the Lizard in a vast territory that boasted a population of less than 20,000, where there were no cities and where the only means of communication lay along the rivers and Indian trails. Yet in 1853 the Methodists had worked out the beginnings of the first university to be founded in the North-West; more than a hundred years later Cornwall, with a population of more than 300,000 still lacks its university. Incredible as it may seem, in 1871 James Matthews of Penzance, a cabinet maker by trade and a Methodist preacher by calling, actually left Penzance for Oregon with his wife and three children on money lent him by a friend in Portland because he was "ambitious for education". The days of the Oregon Trail now over, this new generation followed overland by train to San Francisco and then by boat to Portland, where Matthews soon found work at the bench and in the pulpit, eventually homing to Chinook in Washington, gateway to that glorious peninsula of Long Beach with its twenty-eight miles of unbroken sand and fishing grounds for crab that must have brought back sweet memories of Newlyn and Penzance. Here was the New Cornwall of cliff and rolling seas and lighthouses whose state flower, the exotic rhododendron, is almost symbolic of the Duchy of Cornwall; and from here the elder son, James Thomas Matthews, born near Penzance, rode high into the educational inheritance denied to him in Cornwall. After graduating with distinction at the University of Willamette, he was appointed Principal of Lincoln Grammar School at Salem and then Professor of Mathematics at Willamette. "A gentle little man whose personality overflowed with quiet dry humour and love of humanity and faith in simple Christian living", he brought these qualities to his teaching at a time when his only reward was the gratitude of his students, for salaries were poor. Today the four-roomed cottage in Heamoor (near the inn known as "The Sportsman's Arms" on the main road to Penzance), looking externally much as it did when Matthews was born there in 1864, has no marker to this exceptional teacher of the American academic world; yet at the University of Willamette his memory is kept alive by a stained glass window. Had he stayed in Penzance, he would never have found a place among teachers, not even at the local grammar school, still less at a university, for England then boasted but four. Instead he might have followed his father's trade, become a shop assistant or, worst of all, a miner [1].

Others too from Cornwall discovered their heart's desire in Oregon. The brothers Edward and William Grenfel left Penzance for Wisconsin during the Civil War and then travelled on one of the first transcontinental trains for Portland. They brought out the rest of their family and married girls from two other Cornish families already established on the Columbia River, Prideaux and Francis. A descendant, William A. Grenfel, distinguished himself by being elected in 1957 to the House of Representatives at Salem. Two associated themselves with the brewing industry: William Arthur from St. Austell, where he made the casks for its fine pale ales, settled in Silverton; and Fred Matthews from Penzance, after ridding himself of the taste of Montana's mines, grew hops in the meadows of the valley of the Willamette. A Combellack and a Dunstan from the creeks of Gweek married at Seworgan in 1883, and their Californian-born son became a Professor of Classics at the University of Oregon at Eugene. An Elizabeth Roe, who was reputed to be the daughter of a Cornish physician, married a Dutchman, took to the Oregon Trail and homesteaded in Roseburg. Her family still recalls a few words of a song that echoes the depression in Cornwall in the 1870's: "Give me three grains of corn, mother, to keep me alive this night". And almost as completely forgotten is Richard Cook from Liskeard, a disillusioned miner at Gold Hill in Nevada, who operated one of the first saw-mills at Jacksonville.

Walter Rodda Dry, a small slight man of indomitable energy and wry humour, had a mother who lived near Penzance until she was sixteen and then, with her mother Mary Jane Rodda and a host of relatives from St. Just, emigrated in 1870 to Minnesota. He remembers

---

1 See his autobiography, *Turn Right to Paradise* (Portland: 1942); and *The Oregon Statesman*, 6 June 1942.

her in the flickering light of recollected stories—of her playing over the rocks near the Land's End, of childhood games more English than American in which "good Queen Bess and bluff King Hal" were the stock characters rather than Washington and Lincoln, of ships wrecked on the Cornish coast, and of mines that were worked beneath its seas. Mary Jane Rodda's husband was drowned at sea, and in Minnesota she married again, and of course another Cornishman, Andrew Nichols. Their son Richard Rodda became a miner like his father and worked around Lake Superior until he moved to Terry in South Dakota to manage the Horse Shoe mine; later he was in the employ of the government of Mexico as a consultant. Mary Jane's daughter too discovered to her delight that the American environment could work wondrous spells on unsuspected abilities, for she married a pharmacist and became one herself, a career denied to girls in England for many years to come. Her son, Walter Rodda Dry, superintends like his uncle, but his world is no longer that of the darkness of the mine but of the sightless, for he has been in charge of the Oregon State School for the Blind. More recently, since he retired, he has been expending his energies on the problems of the aged from his bungalow at Manzanita, where he and his German-born wife look out on curving beaches and green headlands that duplicate the coast profile of Cornwall, and meditate on their visit there to search out their origins.

Another family from Penzance, the Switzers of Oregon City, who came to rest by the waters of the Willamette after a lifetime of mining elsewhere, begins with John Terrill and his foolhardy habit of warming a stick of dynamite in the top of his boot to enable it to detonate the quicker, until one day his leg had to be amputated. His son, Charles Frederick, seems to have been more circumspect and preferred the whirling wheels of an engineer in the mines near Golden in Colorado until he and his Welsh wife decided to take their six children to the sunlight of California. It was 1904 and the journey should have been accomplished comfortably by train, but the old pioneer spirit of adventure urged him to attempt it by covered wagon. He never realised that the trails west could be just as arduous as they had been half a century earlier. It proved to be an unforgettable six months' agony across the ashen wastes of Utah to Salt Lake City for they nearly lost a one-year old child, saved only by the luck of having with them a copy of a *Medicology or Home Encyclopaedia of Health*. Wiser counsels now prevailed; they wintered in the Mormon capital, the idea of the journey to California was abandoned and they entrained for Oregon. In Portland work was never easy to find, so the children had to make their contributions to the family budget as soon as they were out of school. The home owed everything to the mother, for she had been trained in the disciplines of the Baptists and was dedicated to fighting the world, the flesh and the devil in the liquor stores and the saloons, fortified by the belief that the best is yet to be in a world whose riches never came even in stray showers.

Only once did the golden fingers of Midas come within inches of touching them when in 1916 Terrill discovered on his property in Oregon City a vein of silica. His hopes ran high for, if the vein flowed deep, Oregon City would have its first mine and his family a considerable fortune, since silica was used in the manufacture of a wide range of products from soap to wood filling. Local bankers and business men backed the project, a small plant was installed of rolls, mills and boilers, and the excited Cornishman was a partner in the Silica King Mines Company. Heavenly choirs of Cousin Jacks long since departed, sang their hymns of encouragement to the sons of Terrill as they dug, ground and sacked the silica for shipment as far afield as Ocean Falls in British Columbia. He ploughed back into the industry all profits, more Californian capital was pumped in, more land was acquired and the old family business was merged into the more grandiose Monarch Mining and Milling Company with Terrill as the general manager. There seemed to be no doubts of a constantly expanding market, especially when tests showed that the silica could be used as a filler in the asphalt industry and as a substitute in Portland cement. There was no scarcity of buyers, among them the University of Oregon. There came cheering reports from the laboratories of the china-clay industry at St. Austell in Cornwall that the Oregon clay showed a silica content of 73.82 compared with 73.57 from Cornwall.

But as the Terrill tunnel penetrated deeper into the hillside, the fickle silica sank from view, investors shook their heads and withdrew their shares, and the family once more was left in the dust to fill the bags themselves. Hopes flickered again when the Metals Extraction and Refining Corporation of Ogden and other companies found other metals in the silica samples, gold processing out at 26 dollars a ton, palladium at 448 dollars, platinum at 126 dollars and iridium at 75 dollars. But Terrill knew from his inborn Cornish experience that trace metals were usually far too expensive to extract, so he continued with his silica digging until the depression of 1929 put a stop even to that. Bedevilled by law-suits, claims and counter-claims by creditors, he died of heart failure in 1947, his middle-age dreams of moderate wealth eroded away among his rusting machines and his derelict warehouses.

East of the Cascades, where the metalled veins of Oregon meet those of Idaho, another Cousin Jack drank the waters of disappointment. Richard Thomas of St. Just, who had been trained at the Camborne School of Mines, in the severe winters of 1902-4 was scratching letters home from the Alps mine, high on Bald Mountain above Central City in Colorado. He complained that "the camp here is still very dull, most of the men leasing and not earning any money", that Telluride, "the liveliest camp in the State" had been shut down by its English capitalists because "the Miners' Union had killed Arthur Collins Super of the Smuggler Union mine", and that thousands were out of work[1]. So he found himself in the ranks of the unemployed, glad to make his grubstake placer mining at Silver Creek and occasional shaft-sinking for a Dutchman.

In the fall of 1904 he struck out hopefully for the Pacific North-West by train from Denver and settled for Wardner in Shoshone County, hardly realising that he had alighted in the most historic mining area of Idaho, where on the Coeur d'Alene River in 1865 the first of the gold-rushes had spent itself. Concerned less with history than work he could not have arrived at a more luckless hour for, as he reports, "the whole country around here is on the bum on account of water and all the mines are being closed down". Two months were spent in a fruitless search for work until finally he was reduced to the humiliating position of having to write to a brother in St. Just for a loan, an unheard-of request in that town. One of his disappointments was that he rarely came across any Cousin Jacks, so it was with a real feeling of relief that he eventually ran to earth Richard Pascoe, the general manager of the Standard mine at Wallace with a lifetime of experience behind him; moreover, his brother was manager of Botallack mine at St. Just. But it was a wet mine and a hazard to health: "three successive nights last week I came off work soaked to the skin and before reaching home all my clothes were frozen fast on me and I couldn't take them off myself".

He then tried the Tiger mine at Burke where "the work seems easy" at the rate of 3.50 dollars a day and where he could save about 40 dollars a month to pay his debts at St. Just. But he never really liked it, for there were no other Cousin Jacks at the mine and "everyone you speak to are strangers". Trade unionism was on the march, "the President of the Union intends to come here and try and do away with the employment system", and labour troubles were in the air. So, sick for a

---

[1] His correspondence is in the possession of Mrs. J. D. Pawlyn of Newlyn, Cornwall who kindly made the letters available for inspection. See also A. C. Todd, 'Cousin Jack in Idaho', *Idaho Yesterdays* (Vol. 8, No. 4, Winter Issue 1964-5).

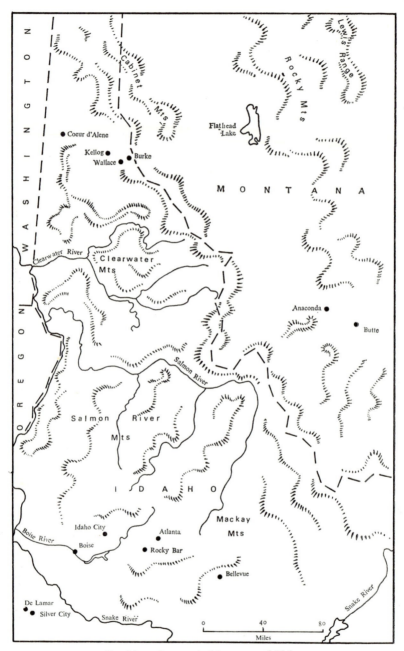

Cornish settlements in Montana and Idaho

sight of the Cornish and nursing a cut along the back of his hand "about two inches long and wide enough to put my little finger in" that a fall of rock from a hanging wall had ploughed, he made a bee-line for Grass Valley. But four years of war then separated the two Cornwalls and Thomas wandered disconsolately from camp to camp until he had saved enough dollars to make the long return home, only to be cheated on the very morning of his departure, for he died in the street of a heart attack.

Richard Thomas, with his banjo on his knee and rivers of hope pulsating through his veins, was a nomad of the plains and mountains in a part of Idaho so bleak and inhospitable that Stephen Douglas had remarked that it could never be settled until the twentieth century[1]. It seems reasonable to suppose that there were some Cousin Jacks among the unnamed thousands who left California in 1855 for the Fraser River and then turned south along the natural extension of the Northern Mines of British Columbia into Idaho Territory four years later. There is no evidence however that they were concerned in the first serious prospecting and important discoveries with E. D. Pierce on the north fork of the Clearwater River in 1860.   Yet a convincing piece of evidence that there were some Cornish in the Boise Basin in 1865 comes from *The Idaho World* of Idaho City of 12 August. It quotes a report from the Republican *San Francisco Flag* concerning the attitudes of Cornish miners in the Mother Lode:

There is a voting precinct in Nevada County, not far from Grass Valley, known as "Allison's Ranch", which gave McClellan 328 votes and Lincoln ONE. The men who cast this vote are SUBTERRANEAN MINERS— BRUTISH AND ILLITERATE CORNISHMEN—who work from one year's end to the other four or five hundred feet underground, and come up with the light of day, only when they are hauled in a car, to their meals . . . THE UNION CAUSE CAN AFFORD TO DISPENSE WITH ALL AID COMING FROM THAT QUARTER.

The Idaho World therefore hotly came to the rescue of all Cousin Jacks in its neighbourhood:

Miners can judge for themselves what a friend they have in the party which makes *The Flag* its official organ. As "the Union cause can afford to dispense with all aid coming from that quarter", we think some of these "subterranean miners", so contemptuously spoken of, will here in Idaho Territory next Monday show that they are not so "brutish and illiterate" that they cannot resent so flagrant and uncalled for an insult.

Idaho Territory, or the "Gem of the Mountains", was created in March 1863 from portions of what was previously known as Washing-

---

[1]   Beal and Wells, *A History of Idaho*, p. 281.

ton, Dakota, Nebraska, Montana and the greater part of Wyoming, with an impossible area of almost 350,000 square miles. Five years later it was reduced to a more manageable 84,800 square miles, but it still had to wait until 1890 before it acquired statehood, so slow was its population growth. Structurely it consists of a huge wedge-shaped tableland, rising gradually from 4,000 to 10,000 ft, "literally crumpled and rolled up, in one continuous series of mountain ranges, fold after fold", Coeur d'Alene and Bitter Root to the north, Rocky and Wahsatch to the south-east and the Owyhee to the south-west, the whole traversed by four major rivers: the Snake, the Clearwater, the Spokane and the Pend d'Oreille, all emptying themselves in the Columbia. Of these the principal was the Snake, "erratic, turbulent, sluggish and deceitful", trapped by the French *voyageurs des bois* for its beaver and serving as the western boundary of the state.

Some of the first Cornish made their way forty miles to the north of the original French settlement of Boise to Idaho City, believed to be at one time the liveliest town between the Mississippi and the Pacific. Originally called Bannock, it was perfectly situated at the confluence of Mores and Elk Creeks, whose beds and gravel bars were loaded with gold when the Grimes' party came upon them in August 1862, and is the heart of the Boise Basin's eighteen square miles of mineral-rich earth that yielded "more gold than ever came out of Alaska", to quote a sign at the entrance to the "city". Today it claims several "firsts" in Idaho: the Masonic Hall, the newspaper building, the Oddfellows' Hall and the Catholic Church, all serving a population of almost 5,000 miners and their families in 1880. It can be approached along the historic road which for several miles is part of the Oregon Trail, still clearly seen driving across the wilderness of Rock River which has cut a canyon through successive age-old lava flows and runs between towering cliffs of columnar basalt. In the canyon the air can be savagely hot and still but today, as you drive towards Mores Creek, relief can be found in the Lucky Peak project that harnesses its waters to irrigate Boise and blows white jets of spume, cooling the air for half a mile around. Towards Idaho City this creek opens out to a wide valley, flanked by the cool shade of pines and wild syringa. Here is visible evidence in abundance of placer, hydraulic and dredger mining: the mounting mounds of boulders, glistening white on the banks in the glare of the sun, placed there by the blistered hands of nameless miners; the earthen banks left by later generations of hydraulic miners; and the shapeless masses once manipulated by the unwieldy dredgers, one still hulked down on the goldless sand.

It is only within the last few years that the ghosts of Idaho City have been exorcised and the spirits of miners long since dead placated by a people now conscious of its responsibilities. In 1958 its Historical Society started an annual fiesta, called "Gold Rush Days", one weekend in June, as a rehearsal for its centennial celebration in 1962 but then discovered it had more important work to do on its pine-covered Boot Hill. *The Idaho Sunday Statesman* of 15 June had complained:

Nobody knows how many hundreds or thousands of persons have been buried on Boot Hill. Scores upon scores of head boards have rotted away; picket fences surrounding the graves have caved in; hundreds of mounds were never even marked; trees, planted as saplings at the head of the graves, now are giant pines; everywhere is rot and decay, weeds and undergrowth, ruin and neglect.

Residents of Idaho City were quick to take the hint and today the receipts from "Gold Rush Days" are used to resurrect the graves and boards of these mining pioneers. But for this work of redemption, lost forever would have been the grave of "James Scaddon of Redruth, Cornwall, England, who died Banner, Idaho, March 22, 1887, aged 49 years"; and that of "Annie Olds, aged two". No one knows what happened to George Mitchell and "the Tregaskis girls" whose names appear in the old store's account book, preserved in the museum that was once the Wells Fargo station, as having "paid in full" for their muslin, satin, holland, hats, shoes and mitts; nor to J. W. Drew, miner, M. T. Pierce, carpenter, and W. E. Rowe, blacksmith, all of whom were working in Idaho City in 1865, according to *The General Directory and Business Guide of the Principal Towns East of the Cascades*.

A more difficult camp to reach, even now, is Rocky Bar. With Atlanta it lies outside the Boise Basin proper, but it was here in May 1863 that the rich quartz discoveries were made that attracted a population of more than 5,000. A year later it was the seat of government of Alturas County, a vast area that included all lands north of the Snake River as far as the Saw-Tooth Mountains. The approach from Boise is sensational and dramatic; and one can only marvel at the feats of endurance and organization that carried the miners to their make-shift cabins on Red Warrior Creek, and transported the ore for shipment along the many river trails. Today, there is first the long haul over the desert along the modern freeway (part of it is a portion of the Oregon Trail that was staked out by the indomitable Bonneville) with the Owyhee Mountains to the right, on which the snow still lingers in July as a cooling thought against the hot blast that sweeps the desert. But at 3,000 ft the desert changes gradually to meadowland that recalls a

Cornish moorland softly undulating towards the sea; but in place of the sea is the Long Tom reservoir, one of a series of inland waterways that brings life to a thirsty Boise and a breathless beauty to the mountains. Another is the Anderson Ranch Dam, the biggest earth dam in the world, fringed with tiers of ponderosa pines, beneath whose blue waters run the miners' trails that eventually cut through defiles of granite to take them through the pass that opens out into Rocky Bar.

In 1960 its population comprised one Yorkshireman from England; in 1964 he was frozen to death, trying to escape south from one of the worst blizzards of the century. Rocky Bar is also the graveyard of a once thriving mining population, though this is difficult to believe since Nature has mellowed it to a beauty that is as timeless and as indefinable as its silence. The stillness has a background of moving water, of bright sunlight on an abandoned drift and of deserted buildings that evoke the twin tragedies that overwhelmed the community: first a fire and then a fatal cloudburst. There remain a crumbling livery stable with room enough to stall forty horses, rusting stamps, giant ore dumps, the shells of cottages and undisturbed dust roads. A wide open space, now covered with a riot of long lush grasses, where stood the courthouse, a hotel, boarding houses and the church, is a reminder that here children were born, played their games and went to school in summers that were an idyll the long day through. And here too their parents died, to be carried over the shallow stony creek to the dark-pined hillside, where today there remain few headboards and no sign of the Cornish.

The earliest accounts of mining operations at Rocky Bar appear in the reports of special agents who were despatched over the entire West by the United States Treasury Department, published under the title of *Mineral Resources of the United States*. In 1868 J. Ross Browne reported that Rocky Bar was a small compact district of many veins of limited compass that had been the scene of many enterprises by eastern capitalists, all of whom concluded that the mines were not a really sound economic proposition because of their inaccessibility, all freight having to be transported on pack animals. Even when a wagon road was constructed in 1868, the cost of carrying machinery and provisions remained far too high and the main cost of production, the wages of the miners themselves, became prohibitive[1]. Agents continued to be sent every year from 1870 to 1876, and all their reports proved most discouraging. For the year ending 30 June 1873 Rocky Bar could only manage to send down 35,000 dollars' worth of gold to Boise because of

[1] *The Idaho Weekly Statesman*, 18 May 1869.

the problems of milling. The mill of the Valley Gold Mining Company possessed only one Dodge crusher with a pair of Cornish rolls, and most of the work had to be performed in a primitive Mexican stone arrastra, consisting of circular pits, lined at the bottom with smooth stones, in which the ore was flung; it was crushed by a large boulder that was dragged round the pit by horses.

The chief mine was the Ida-Elmore, and in 1875 the special agents were reporting that it was producing "the most flattering" results, but that its owners, the Pittsburg Company, could make it work even better "if capital were judiciously expended". It seems that the improvement coincided with the arrival of its new superintendent, William Rowe of Redruth. When he first reached Idaho is not clear, but it is known that in 1874 he was the Master of the Boise Shoshone Lodge and that in 1875 he transferred his membership to Alturas Lodge No. 12 at Rocky Bar. Apparently he was so charmed by the place that in 1876 he actually returned to Cornwall to claim a wife. *The Idaho Tri-Weekly Statesman* of 12 July 1877 was happy to report: "Our old friend and resident of Rocky Bar has returned after two years' absence, accompanied by a most accomplished and beautiful wife together with his sister, a charming young lady. Welcome back, you and yours, Brother Rowe". The names of the Cousin Jacks who roared their fraternal greetings have been lost in the fire of 1892, but two lie in Boise cemetery beside Brother Rowe; a Blewett and Thomas Davey, born in 1839 at "Mount Hawke, England". Ask anyone in England the whereabouts of Mount Hawke and they will not know, but in 1903 it was so well known in Boise that the inclusion of "Cornwall" was not even necessary.

Others may be found in the pages of the 1880 census returns for Alturas County: William Jennings (22), Able Rowe (29), Hercules Prisk (32), John Thomas (28) from Brazil and his two uncles, one of whom was paralyzed, Charles Davey (37), Thomas Kitto (37) and Samson Tonkin (28). Remembered even for their physical prowess in the official returns, they seem to have been led by John Tonkin who had married a Welsh girl in New York and served as an Indian scout; John Andrewartha, the Methodist minister from Georgia, who married, baptized and buried them; and Charles Puchenhorn (sic) Davey, the foreman of Ida-Elmore's rival, the Buffalo silver mine.

At Rocky Bar, in the tiny chapel among the pines, Davey married Louisa Jane Rowe, the sister of William Rowe, who bore him seven children before she was thirty-three. Two were destined to die in Nevada, one killed outright through mistaken identity and the other of "black measles" at Goldfield. "An old and experienced miner of many year's practice, whose work cannot but prove satisfactory" to the Buffalo Company, it was reckoned that under Davey's direction "by the time the furnace is in full blast and the sound of the new mills and stamps are heard echoing by the surrounding mountains, Alturas will be equal to that of Virginia and the Comstock mines"[1]. With the encouraging news that Davey and another Cousin Jack, R. Scovell, had found a very promising quartz "well in free gold and improving as their shafts extend downwards", a new furnace was blown in on 17 March 1877[2] and once more the Cousin Jacks dreamed of their silver dollars rolling into the banks down at Boise.

As at New Almaden, so at Atlanta and Rocky Bar, the Cornish miners and their families were forced back upon themselves to provide their own amusements and social security. Alturas Lodge No. 12 of the Independent Order of Oddfellows was, under its Cornish officers, as lively as any between the Mississippi and the Golden Gate. John Davy, R. O. Hoskins and William Rowe organized a "grand Masonic Ball", described as "one of the grandest affairs ever held at the Bar" for which "elegant printed invitations and cards with programmes of dancing have been gotten out"[3]. Almost twenty years later, after the disastrous fire of 1892 that started one September day in the laundry room of the Alturas Hotel and caused damage to the extent of 75,000 dollars, John and Willie Davy were the civilized socialites of this rocky principality, arranging their masquerade balls in the same way as their fathers. Visitors were delighted by this seeming paradise that the Cornish had carved on the side of a mountain where "Miss Stacey and Miss Rowe are charming and entertaining young ladies who fascinate the young and interest the old". But it was not always a summer's day, for life could only be prolonged while the ore lasted and while the Cornish pumps lowered the water level in the Elmore shaft[4]. The winters could be bitterly bleak. Some Cornish filled in their time by breeding dogs and one, in St. Bernard fashion, rescued the son of blacksmith Davy by "uncovering the boy's head long before human help reached him"[5]. Some, like W. P. Rowe and Tom Kitto who owned a partnership in the Mountain Buck claim just outside the Bar, might volunteer to tackle the treacherous descent to Boise with a patient for the hospital or "to help straighten out a senatorial tangle"[6]. Some, like Frank

---

[1] *The Idaho Tri-Weekly Statesman,* 13 July 1876.
[2] Ibid., 22 March 1877.
[3] Ibid., 12 June 1875.
[4] Ibid., 24 July 1895.
[5] Ibid., 23 January 1892.
[6] Ibid., 5 January, 2 February 1895.

Mitchell and Will Davy, would snowshoe twelve miles over the mountains to Trinity Lake to chisel through its five feet of ice so that the mining garrison could break their winter fast with fresh fish[1]. And others would elect to keep open the school, like Charles Davey who was a trustee of the Rocky Bar School District[2].

No doubt the parents, as well as the children, attended school if they were sensible, as apparently Henry William Warren of Camborne was. He had little choice in the matter for, after many years in Cripple Creek, he could still neither read nor write until he married the school teacher at Rocky Bar. Laura Emerson was the daughter of a Scots farmer from Alleghany County in New York who emigrated to the Cornish backwood country around Whitewater in Wisconsin, carrying around with her an uncompromising opposition to drink. For a time, while teaching the children of a Jewish merchant in Boise, she got herself made president of the Anti-Saloon League of Idaho; and then, to the consternation of the miners of Rocky Bar, appeared in their midst. Yet she saved miners like Warren from educational atrophy, putting them through the hoops of their letters and numbers and binding around them the straitjacket of total abstinence from drinking and smoking. With his pockets packed with pencils of lead and dynamite, Warren blasted his rock and composed sentences for his own edification and willingly handed over his eight children to his wife to educate. One became an attorney, another a real estate broker, another an underwriter, and the remainder farmers; but they all enjoyed the benefits of a university education, a blessing which their father, when he died in 1938, must have regarded as well nigh incredible, so rapid had been the transition from illiteracy. The Cousin Jacks will never need reminding of the figure of the school ma'am and her schoolhouse, or chy-an-schol, astride the mining frontier.

Equally successful has been the Rowe family, though their history reveals how narrow was the gap between life and death in the chance world of mining. William Rowe had returned to Rocky Bar in 1876 with his young wife from Redruth in the hope that the altitude might prolong his life, for he was, at forty-four, an old man with rotting and disintegrating lungs. "A good man, a zealous Christian and a kind husband performing all his duties as conscience directed"[3], he lasted only four more years, leaving his shocked widow with a girl only one

1 Ibid., 12 April 1895.
2 Ibid., March 1892.
3 *The Idaho Statesman*, 24 April 1880.

year old. There was another child on the way and she had to seek shelter in one of the boarding houses until another Cousin Jack married her. Five years later she too died; and her second husband inexplicably disappeared into the outback of Australia. The two children, now orphans, were first cared for in Rocky Bar by the local lodge of the Free Masons, then by a merchant in Boise, and finally by the family of General Lafayette Cartee, the Pennsylvania-born mathematician who went to Idaho in 1860 to become its first Surveyor-General, and whose frame house on Grove Street was for many years one of the stately mansions of the city.

A hundred miles west of Boise in a high mountain valley that tops almost 6,000 ft clung three other camps that attracted the Cornish: Silver City, De Lamar and Dewey. Owyhee County (a phonetic name given to the river that flows west of Boise by a Hudson's Bay trapper in memory of his two partners from Hawaii) was a flourishing mining area from 1863 to 1875. Cousin Jacks were employed on the Morning Star and the Oro Fino lodes at the War Eagle, the Chariot and the Poorman mines, and danced their nights away at the Tregaskis Hall when Silver City was designated the seat of government in 1866[1]. An early note about one of them comes from *The Arizona Miner* of 27 April 1872, which quoted from *The Owyhee Avalanche* the news of an "uprising" of 300 miners, mostly from the Mahogany mine, as a protest against their tyrannical Cornish foreman, John Jewell:

The men held a meeting and notified him to leave, but the acting superintendent of the mine backed Jewell and sent up from town 40 armed men and two howitzers to keep the peace. Then the miners from four neighbouring mines got together and agreed not to fight and not to go to work as long as Jewell remained at the mine. Jewell vacated the premises.

By the end of the century the Cornish tide had swept in at the flood as Dewey and De Lamar sprang into life through the enterprises of their founders. Captain De Lamar, having bought the Wilson mine for 10,000 dollars, developed the property by systematic tunnelling, for which work the Cousin Jacks were in great demand; and Colonel William H. Dewey, a bankrupt when he arrived in the gulches with his partner Michael Jordan, built roads and laid out the townsites of Silver City and Dewey so well that in eight years the population of Owyhee County more than doubled.

In 1898 newcomers to Silver City, and there were about 1000 of them, were surprised to find the climate far more attractive than they

---

[1] *The Owyhee Avalanche*, 25 November 1866.

expected; lying in a canyon at the head-waters of Jordan Creek where the altitude stood at 6,300 ft, and flanked by the 8,000 ft Eagle and Florida Mountains, in summer it was hardly ever too hot by day and the nights were invariably cool, while in winter the snows were never too severe. Cousin Jacks, still alive, enthuse about the call of these wilds that surrounded the Black Jack and Trade Dollar mines and the isolation that forced them to devote so much of their spare time to the lodge work of their masonic orders. The 1898 official *Directory* of Silver City lists about forty of them and adds a few biographical notes. James Pascoe joined the ranks of the "immortals" as deputy-sheriff and constable. Thomas Drew shone as one of the early presidents of the Silver City Union No. 6 of the Western Federation of Miners. James Rodda, Fred Tyacke, John Pascoe, Joseph Prout and Archie Warren sat on the committee of trustees of the De Lamar branch. William Toy presided over the branch at Fairview with many years of mining to his credit at the old Cliff mine on the Keweenaw Peninsula and in Virginia City before settling down to a sensible retirement raising stock and ranching. Of an older vintage was John Mitchell; taken as a baby from Truro by his parents to Milwaukee in Wisconsin, he was grubbing in the gulches of Silver City when Toy was still scraping the copper in Cornwall. He too finished out his time ranching and running a successful livery business in Silver City and Mountain Home. Then there was Simon Harris, born in Wisconsin and a lead miner like his father; he rose to be foreman of the Black Jack and a magistrate too. Another Justice of the Peace came from St. Ives. Richard Haws, who emigrated when he was eighteen at the close of the Civil War, mined in New Jersey and Connecticut, crossed to Grass Valley by way of Panama, and then the Sierras to Virginia City, finally trudging the Bonneville Desert to Idaho, where he staked his all on some shares in the After-thought mine and won. "Happy as a sunflower" when he married Phillipa Edwards of Grass Valley, he invested all his winnings in an eating-house but, both of them being Methodist, it was a temperance establishment.

Another Cornish casualty of the post-war depression of 1866 was Thomas Barnes of Wall in the parish of Gwinear who, at the age of sixteen, pitched himself and his father on to the sunbaked wastes around Bingham in Utah. For thirty-four years he mined all over America, never once returning to Cornwall until 1901 when he set out on a pilgrimage to find himself a wife. Choosing Frances Jane Knapp, he promptly whirled her off to Silver City where he was a shift-boss. The return was perhaps a mistake, for the mines were closing down, while his wife found the altitude and the snows a trial to her nerves and eyes; so back they came to Cornwall with their two American children. The son was later to reverse the process by settling in Detroit with his Cornish wife from Heamoor for she had an uncle there, Sam Polglase, whose mining career spanned from Upper Michigan to New Mexico. The daughter elected to stay in Cornwall, claiming dual nationality.

Some Cousin Jacks of substance shrewdly calculated that they would be best remembered by posterity if they appeared in the official histories of Idaho, no doubt in return for a subscription. John Kent appears in one writtten by Hiram T. French in 1914 and is described as a blacksmith who, after only one year in Boise, bought the Idaho Carriage Company in 1905; he was undoubtedly "a Republican of the independent class which allows no party traditions, however noble, to blind the voter to the real issues of direct moral influence". Mark Wilton and Thomas Richard Faull find a place in another released by the S. J. Clarke Publishing Company in 1920. Wilton, after ten years of mining in South Dakota and Montana, had retired to a twelve-acre ranch, the "splendid appearance" of which "indicated his life of untiring activity intelligently directed". Faull, also ranching sixty acres at Emmet, "one of Idaho's pioneers . . . throughout the period of its early epocal developments", apparently "has never had occasion to regret having formulated the plan that brought him to the new world, for he has found the opportunities which he sought and in their utilization has made steady advancement". Finally Luke Williams makes his bow in Byron Defenbach's *Idaho, A History of the Gem State from Prehistoric to Present Day*, published in 1933 by the American Historical Society. Phrases like "a man of sturdy and industrious disposition" and "a hard worker in everything he undertakes" in his case are no mere formality. Coming from a Cornish family "liberal in politics" and "Wesley Methodist Church", Butte made him. By 1905 he had actually saved 5,000 dollars, turned his back on mining forever and bought a ranch at Rupert. Americans in Idaho regarded him highly, making him chairman of the Rural Electric Company, of the Mutual Telephone Company and of the local wool pool; more than that, because "he enjoys politics", they sent him to the State Legislature twice, in 1923 and 1925.

In the trout-teeming country around Bellevue, where the Galena and the Minnie Moore mines were found, in 1960 lived Nick Werry, who had notched his ninetieth year to Heaven. St. Austell bore him in its vicious cradle of clay pits and wilderness of dusty poverty, where he and his brothers "ate their peck of dirt early in life when their mother would take them out into the fields and put them down in the hedge while she worked". His father had a brother in Nevada and, as so often happened, he was the means of rescuing the others from the trap of St. Austell. They crossed the Atlantic in the *Great Eastern* when Nick Werry was five months old, spent three years in the coalfields of Maryland and Ohio (their mother often wanting to return to Cornwall) and in 1874 reached Ruby Hill in Nevada by train and Concord stage, the boys riding shot-gun with their father and the mother bouncing inside with their infant daughter born in Ohio. At Ruby Hill she almost turned for home again when she saw her relatives actually living in tents and realized that she would have to do the same.

Nine years they lived at Ruby Hill, the two patriarchs following the glittering silver and lead veins and their children the lessons to be learned from the life outside. To make a little more money the men would lease a piece of ground, break the "muck" and expect the children to fill the buckets. There was no piped water, and the children carried it all the way from the mine in two five-gallon cans suspended from a yoke across their shoulders. With no thought of the danger, they would blast out the stumps of trees with Giant Powder for fuel, and take them home in their thin Cornish wheelbarrows. They were unaware of the tragedy that could strike without warning: "we lived on the slope of a hill and one day mother was on the trail that went down the hill when a man passed her with an axe on his shoulder, who wished her good morning, and went straight to a boarding house and killed a waitress". Their mother worked in a boarding house until they had accumulated capital to buy one of their own, employing Chinese servants and feeding forty miners a day. Yet somehow she found time to raise eleven children.

When Nick was about twelve, the two families separated and his family opted for Bellevue in Idaho, leaving the others to struggle in Russel Gulch, Colorado, where the winds from Caribou blew so strongly that the shaft house had to be anchored with steel cables. Bellevue seemed a miner's paradise above ground for a river ran the entire length of the valley, where cattle and horses could be raised and most crops grown. "Here", says Werry, "I grew to manhood, became a hard-rock miner and later ran sheep."

For the first time in his life, his father homesteaded twenty-five acres and the boys were set the task of forking and spading. Below their property was the six-year old Minnie Moore mine, bought by Dent, Palmer and Company from Daniel Scribner for less than 500,000 dollars. The leading Wood River silver producer, its reserves were estimated to be nearly 700,000 dollars, but later it topped 8,000,000. The ore was formed in well-defined veins of silver, lead and a little gold that ran east and west, varying from one to thirteen feet in width, with hanging walls of granite and the foot walls of limestone. It was here that Nick Werry "qualified", working ten hours a day for 3 to 4 dollars. As he says: "In the early days it took a year or so to make a good miner —what I mean is, you could take your place in doing anything round a mine, do timbering, take care of the pumps, do shaft work, and know how to take out and save the first-class ore".

Perhaps one of the very last to work over the old trails in the old manner was Richard Terrill who, as recently as 1928, was employed in one of the oldest of Cornwall's mines, Wheal Busy, near Chacewater. The 1929 depression drove him out to Timmins, Ontario, but he had not expected to run into another in America. Nevertheless, he refused to submit to defeat and return home, as many were forced to do, and swung his way from mine to mine through eight states, remitting home £10 every month. He timbered the Scotia shaft in Grass Valley; he crossed the Sierras and worked at Mountain City and Battle Mountain; he turned south down the thick spine of the Rockies into New Mexico, where his tunnelling skills loosened the desert sand for the proving grounds of the very first of the ballistic missiles. He followed the old Santa Fé trail to Tucson and the Tiger mine, where all the bosses were St. Justers; he turned north to the Belmont mine at Butte, then south again to the heat of Tooele in Utah, and finally north again to the Jewell Shaft of the Sunshine mine at Bunker Hill, where two of his uncles, R. D. Jenkins of Redruth and I. J. Sims of Carharrack, were already in occupation.

The Bunker Hill mine, discovered in 1883 by Noah Kellog, is now one of the largest silver and lead pits in the world and it is here that the last of the Cousin Jacks in Idaho have congregated: Sidney Caddy from Illogan Highway, William Furze from Redruth, Herbert Angove from Camborne, and others. In the northern panhandle among the mountains of Coeur d'Alene, reputed to be one of the eight regions in the world that have each produced more than one billion dollars of mineral wealth, this is the place where the first cross-cuts were made by the prospectors from the Fraser River in the 1850's. And this is the place too where, for the Cousin Jacks of the Pacific North-West, their frontier ends. They preferred the mountains and the forests to the twenty acres of land they might buy in Cornwall with their savings, as some did who came home to restore their damaged lungs in a climate whose dampness soon killed them.

VIII

BACK TO COPPER: UTAH, MONTANA AND ARIZONA

FOR the Cousin Jacks everywhere on the North American continent, their frontier halts before the depths of the vast copper amphitheatres of Utah, Montana and Arizona. Thus the Cornish experience that changed the face of American mining begins with copper and ends with copper; and this perhaps is as it should be, for copper was the metal they had known for centuries.

Utah at first held out no attractions for miners; more than half of it was rock, desert and wilderness; and even today most of its million Mormons and Gentiles live in a narrow fertile belt that is no more than 100 miles across and sprawls 400 miles from Idaho south to Arizona. The miners hurried through Parley's, Cottonwood and Echo Canyons as fast as their mules and oxen would allow them, anxious to leave the salt wastes behind in their endeavour to reach California on time. Those who stayed were searching for Zion in a place where they would be free of the red hand of persecution; and among the first was numbered one Cornish family that became a pillar of fire across the desert to Salt Lake City.

John Rowe Moyle, who was born in Wendron in 1808, had moved his family to Plymouth during the depressions of the 1840's in order to find employment on the building of its breakwater, for he was an expert in the cutting of stone, as was his son James, born at Rosmellen, Cornwall, in 1835[1]. In Plymouth they were converted by a Mormon missionary and, when the time came for them to "gather to Zion", it was decided that James, then eighteen years old, should borrow from the Mormon Emigration Fund and prepare the way for the rest of the family. So in March 1854 he embarked at Liverpool in the company of about 400 other English Saints. The journey occupied six months: New Orleans, the slow ascent of the Mississippi (on a boat that was infested with cholera and killed off more than a third of its passengers) and then westwards from St. Louis in a caravan of fifty wagons. He could not have arrived at Salt Lake City at a more exciting time, for 1854 was the year when the walls of the Temple, 9 ft thick, were be-

1 See Gordon B. Hinckley, *James Henry Moyle, The Story of a Distinguished American and an Honoured Churchman* (Salt Lake City: 1951).

233

ginning to rise from their massive foundations; and he was a worker in stone.

For two years he feathered, split and dressed the granite blocks, and regularly subscribed to the Emigration Fund so that his family might make the pilgrimage west. They unfortunately were denied the luxury of wagons, and took part in that epic walk, pushing their handcarts more than 1000 miles from Iowa City for 110 days through the summer heat of 1856. The bleaching sun, torrential rain and hurricanes of wind, shortage of food and water and the unaccustomed altitude all took their toll. When they staggered into Salt Lake City on 26 September 1856, to be greeted by Brigham Young and a brass band, young James Moyle could barely recognise his father, his hands looking like the claws of an eagle from their persistent clutching at the handles of his cart.

Almost a year later came the call to bear arms in defence of the faith. Persecution and isolation they knew had been the lot of the Saints for many years, but they were ignorant of the series of crises building up ever since 1850, when Congress refused to create the State of Deseret. The arrangement by which Young as Territorial Governor was assisted by three Mormons and three Gentiles never worked and only led to rumours that the Mormons intended to set up an independent state. When the federal government therefore despatched 15,000 troops to the Holy City, Brigham Young took up the challenge, cleared it of the old, the women and the children, and ordered 1000 militiamen to block the entrance to Echo Canyon. Among them was enrolled James Moyle as a sergeant in the Nauvoo Legion. All through the winter of 1857-8, with the taste of the infamous Mountain Meadows massacre still in their mouths, the Moyles sweated out their fears of the battle that would come with the spring. But they were spared another "Buchanan's blunder" through the tact of Colonels Kane and Young; federal troops were allowed to pass unmolested through the city; and Utah remained Mormon territory. So the refugees came back to join their families, and among them was James Moyle's wife who, shortly afterwards in September 1858, gave birth to their first child, destined to become one of the greatest of the Saints.

James Henry Moyle, "born in an isolated western community, in an atmosphere of oppression", later went as a missionary to Carolina and Georgia, served as a county attorney after graduating at Ann Arbor (the first Mormon to do so), and during the Wilson Administration was acting Secretary of the Treasury. Later Franklin D. Roosevelt appointed him Commissioner of Customs. His son, Henry D. Moyle, is hardly less distinguished for his services to Church, State and Nation. Ordained as an Elder in 1947, elected in 1953 a member of the Council of the Twelve Apostles after a law training at the universities of Utah, Chicago, Harvard and Freiburg, he is now First Councillor to the President of his Church. During World War II he was a member of the Petroleum Industry War Council, served on the National Petroleum Council and finally was appointed to be Director of the American Petroleum Institute[1].

Both father and son derived their energies and their quite outstanding sense of responsibility and service from young pioneer James, the stonecutter. In time, James became one of the chief building contractors in Salt Lake City and was in continual demand in those early days for the construction of stores, warehouses, houses and bridges for the railroad in Weber Canyon. His chief title to regard now is that he supervised the stonework on the Temple itself. In a different way his father has also left his mason's mark among the Mormons for, when he was moved in the crisis of 1858 to Mountainville (today called Alpine) along with the aged, the infirm, the women and the children, he instinctively fashioned out of the granite mountain a "cave" to shelter them from the federal troops which also served as a "castle" to protect them from Indians. Two storeys high, its walls two feet thick and stressed with natural clay, it seems to have been a perfect piece of military architecture, a miniature bastion in the likeness of one of the towers of the Pendennis Castle he knew and admired in Cornwall. It still stands as a memorial to his skill by the side of a chapel he built in 1863[2].

Almost seventy years later, yet another Cornishman was bound for the land of promise, though more by accident than design: Charles Thomas of Paul, whose father also was a stone-mason, his special craft being that of repairing the walls of lighthouses. On the *Franconia* he met Ethel Dixon from Birmingham, a Mormon en route to Zion. This must have been one of the most unfortunate times to emigrate, with the Great Depression just around the corner; consequently, after a marriage which involved accepting conversion to the faith of his wife, he was unable to find work in 1930 and sent her back to Salt Lake City to await the birth of her child, himself following on foot in the manner of the handcart pioneers. He found that the rigours of the route had not changed in the least and that he could not even share the companionship of those well-organized Mormon caravans. Walking the main streets

1   *The Millenial Star* (Vol. 115, No. 11), November 1953.
2   See *The Deseret News*, 5 July 1963.

and high roads in the heat that rose from the parched earth, sleeping under the stars or in the rain in unfamiliar canyons, nervously 'riding the rods' on freight trains, prodded on by the police and tormented by sage brush country that yields no shade, he arrived in Salt Lake City two weeks before his child was born.

There he entered upon his spiritual inheritance, though he never completely accepted its discipline and dogma, even when he was saddled with missionary responsibilities and was elected an elder for twelve years. Finally he and his wife made a clean break and joined the Baptist Church in southern California, keeping alive their memories of Cornwall through a Cornish Association they helped to found. They were by no means the first to be daunted by the iron standards of the Saints. Even James Moyle was tempted to join the stream of emigrants who regularly passed through Salt Lake City bound for the Mother Lode. No word of the Lord seems to have decided him, but a stone tossed in the air: when it landed with its dry side uppermost, the pre-determined sign that he should stay, he confesses that he was "a little disappointed".

But the good Lord had already prepared riches, not only in Heaven, but on earth too, for the new state was soon to be a beehive for the mining fraternity. The first prospectors were the troops of the Third Californian Infantry, stationed in Utah during the Civil War who, in their long bouts of boredom, found silver and lead in Bingham Canyon and discovered the old Jordan claim, the first to be recorded in Utah. Before the century was out, there were to be many other sensational discoveries, notably in Park City and Ophir; but Bingham, famed later for its colossal pit of copper, together with Lark adjoining it, constitutes the cradle of Utah's mining industry.

Lark, at the foot of the Shining Mountains of the Oquirrh Range, and on the edge of the Bonneville Desert, the dried bed of a lake that once stretched into Idaho and whose fossil beaches are today clearly identifiable, claims no romantic associations with the birds of the meadow, for it was named after a prospector employed by the Rio Grande Railroad Company which laid an ore line from Salt Lake City, about thirty miles away. Jammed against the massive wall of rock that is Bingham, Lark is the little Cornwall of the desert, even boasting a "pub" appropriately called the Drift Inn. Here at one time lived George Reynolds from St. Agnes, owner and editor of *The Bingham Bulletin*, who achieved local immortality by installing street lighting in such a way that the burning lamps spelled out "Lark". Today you may visit Harry Williams from Stithians, who has left the dry diggings of South

Africa in time to avoid the strangling cough that presages an early death; the widowed Mrs. Prowse in her frame house amid a tangle of sage brush on the desert, daughter of a sea captain and married in the church of St. Mary-the-Virgin in Penzance to a miner from Sancreed who was killed shortly after their arrival at Bingham; and the Thomases from Perranporth and Chacewater.

Bert Thomas, too slight and slender it might seem to wield the hammer and drill, comes from Trevellas in the parish of St. Agnes, attended the village school and then naturally went into that local mine at Bolingey. But to work overseas was part of the familiar family experience, for his father, who taught himself to read and write, had travelled South Africa, returning home every two or three years, while a brother had been tempted away to Australia. Nor were the financial inducements to stay at Bolingey compelling, a labourer making as little as eighteen shillings a week. His departure in 1910 was typically Cornish in its casualness for he left his home with every intention of reaching out for Koolgardie but, at Truro railway station, he met a group of miners bound for America and, liking their company, joined them. The cost of the fare, he remembers, was about five guineas for a second-class berth on the *Philadelphia* from Southampton. In New York at the Cornish Arms Hotel, managed with indifferent success by emigrants from St. Austell, he met a man named Kissel who suggested they should take a chance in Bisbee, simply because he "knew someone" there. Completely ignorant of working conditions in Arizona, Thomas went smithing at the Copper Queen in temperatures well over 100°F and in a very short time joined the army of "10-day" Jacks who, blowing their jobs, would make off for near-by Globe or far-off Butte. He elected to try Park City in Utah where the Cornish were blasting out the lead and silver from the Little Bell, the Daly West and the inevitable Silver King; there he married the girl from Chacewater.

In 1917 the pair walked out from their mountain gulches to live in Salt Lake City, he to sharpen his bits at the Garfield smelter until a strike cost him his job and he journeyed east to Sego among the Book Cliffs that dominate the desert wastes where the Green River struggles for survival. In the torrid heat he hacked out coal and lime from the Grantsville quarries in the Oquirrh Mountains, all the time reading correspondence courses to improve his knowledge of blacksmithing and forging metal. At last in 1922 he qualified, earning the respectable wage of 17.50 dollars a day, first with the Utah Copper Company in Bingham and then with the United States Smelting, Refining and Milling Company at Lark. Today in his retirement, he scans from his company house the desert, Salt Lake City and the 12,000 ft peaks of the Wasatch Mountains, just as at one time he had looked on the Atlantic from the sand dunes of Perranporth and the sand-drowned church of St. Piran.

The thunder of surf on the Cornish beach and the white-crested plumes of spray are here transformed into the detonation of explosive charges and the spirals of white smoke in the near-by Bingham pit. This "mountain that has become a hole in the ground" is perhaps the largest single mining project in the world: dynamite has blasted away the peaks of the Shining Mountains and gargantuan shovels bite off fifteen tons of rock in one mouthful, the endless mastication resulting in fifteen billion pounds of copper in just over fifty years. For the Cousin Jacks from Lark the approach is by way of the spectacular Bingham tunnel which creeps four and a half miles into the mountain; for the visitor it is by way of the well-designed new town of Copperton that has been built by the Kennecott Copper Company for its employees who have resettled from the doomed town of Bingham that is being slowly killed by the pit. From the rim the human eye fails completely in its judgment of distance and size, for this pit, cut by Cousin Jacks to an oval as neatly shaped as the egg of a hen, is large enough to accommodate nine liners of the size of the *Queen Mary*; it measures two miles across at the top and is half a mile deep. The oval has been achieved by cutting steps that are 65 ft deep and almost 75 ft high. Yet for every single foot that the pit sinks, it must extend laterally three feet; ironically the time has come when it must engulf the very canyon that gave it life. Bingham Canyon, where once Cornishmen at the Copperfield and Highland Boy mines wrestled for a livelihood and their beam-engines nodded their languorous length day and night, now dozes away its doom. The shutters are drawn and the saloons closed as the Kennecott Copper Company buys up the property to cover it with the waste and rubble of yesterday that the riches of tomorrow may be found.

Strange as it may seem, at Bingham it is progress and prosperity that are the killers of communities and not the cupidity of the capitalist who all too often transfers his interests elsewhere and leaves a trail of human wreckage behind him. This happened in the Shining Mountains along the Pony Express and the Overland Stage trails, where camps like Mercury and Ophir were reduced to a whisper of ghostly shacks when men like Marcus Daly and W. A. Clark struck it rich and then moved themselves to Butte. With unconscious symbolism Daly called his new hole in the ground there Anaconda, scarcely aware that he was of the race of

The chapel, as seen today, that the Cornish stone-mason, John Rowe Moyle of Wendron, built at Alpine, Utah, in 1863, not far from the Indian fort he erected to shelter his family during the invasion of the Mormon capital by federal troops in 1858. [Deseret News]

The livery stable at Globe, Arizona, that belonged to the Pascoes, a well-known Cornish family. [State Archives, Phoenix, Arizona]

The very first bank in Miami, Arizona. Its first cashier, clerk and janitor was
W. J. Ellery of Redruth. It measured ten feet by twenty and the rear portion was
occupied by a doctor. The more permanent building that followed is now the
Town Hall. [W. J. Ellery]

The Old Dominion smelter at Globe, Arizona, familiar to thousands of Cousin
Jacks who went there to mine its copper. [State Archives, Phoenix, Arizona]

mining serpents that would strangle the workers of Utah, Mormon and gentile alike, in their death struggle against tycoons like George Hearst, who found a fortune in the Ontario mine. But as far as they were concerned, men were expendable in the Park Cities of the gulches, where many a hardrock miner was caught in a collapse of rotting timbers and was dug out dead from beneath a mass of rock and gob. Some Cornish had been known to have been entombed for more than a week, their nerves stretched to snapping point by the hostile ticking of their watches which, in the cold blackness 400 ft below daylight, sounded "as loud as a boiler factory in which a drum corps was practising".[1]

The Cornish were astride the mountains of Butte long before Daly and Clark arrived, wrestling for survival with Welsh, Finns, Swedes, Italians, Austrians, Slavs, Mexicans, Filipinos and, above all, Irish. Culturally therefore it was an area rich to catch the ear of singers and writers about folk-lore, like Wayland B. Hand.[2] The Cornish lamentably had no poets of their own, so they have to be contented with brief appearances in the ballads of the Irish, whose minstrels came straight off the streets of Dublin. Steve Hogan, in *In The Dry* notices the mood of confession that comes over the Cousin Jack in the locker room as he prepares for the descent among

> Half-clad miners chewing the rag,
> This one is a thinker, that one a wag.
> A swarthy Hunk sits and grins
> As a shameless Cousin talks about his sins.

Bill Burke praises their fine bearing in the parades of 23 April, St. George's Day, when five bands would be playing "with Sammy Treloar at their head"; and their physical prowess in the festivities of 13 June, Miners' Union Day, for "the Cornishmen of that era prided themselves, not only on their ability as miners, but in keeping in tip-top shape at all times"[3]. Then again Walt Holliday, in the ballad *That's Different,* draws attention to the Cornish reluctance to undertake the common labouring tasks:

> He can take the eighty penny strike
> And bend it in his hand—

1 Frank A. Crampton, *Deep Enough* (Denver: 1956), pp. 101-117.
2 See his articles in *The California Folklore Quarterly*, January and April 1946, on which the early paragraphs about Butte are based.
3 *The Montana Standard*, 20 April 1958.

> The strongest little Cousin Jack
> That ever struck the land.
> Though when it comes to loading rock
> He will not do it—nay
> He wouldn't load a car of ore
> Not in a twelve-hour day.
> In wrestling down upon the mat
> This Cousin's the best bet
> But the fellow who can make him work
> Has not seen daylight yet.

The hatred of the Irish for the Cornish is understandable though it cannot be explained entirely in terms of Roman Catholic and Methodist bigotry. Celtic temperaments clashed when the exhausted Irish labourer found himself face to face with the acquired skills of centuries. Most of them had never seen a mine until they arrived in Butte and many were taken straight off the train to a basement in the Florence Hotel for preliminary training. Feuds could be sparked off by a mere chance remark or epithet, an Irishman being called a "chaw" and a Cornishman a "petticoat". William Vine of Denver remembered how a Cornish saloon keeper on 4 July 1895 decorated his bar with British red, white and blue bunting, whereupon the Irish threw sticks of dynamite through the window and killed several bystanders. Sometimes this rivalry could be turned to good account by managers setting them against each other to step up production. Occasionally, in the face of other nationalities, they would close ranks, the Cornish celebrating St. Patrick's Day and the Irish St. George's Day; but all too often their mutual bickering betrayed them into the hands of political tycoons so that the Cornish, never interested in the politics of power, were gradually submerged.

An exception here should be made in the case of Absalom Francis Bray, a merchant's clerk in Truro who, in 1876, entered upon his American life as a "navvy" on the Texan railroads and then as a government contractor for the construction of levees on the Mississippi. The work and the humidity however drained away his health so, in 1895, he journeyed to Butte where, with less than 3,000 dollars, he started a small grocery business which was to grow into one of the largest concerns in the whole of Montana. Like most Cornish he was a Republican, but of more than usual ability, for he was elected twice to the Montana Legislature and was once its Speaker.

Had there emerged more Brays in Butte the Cornish might not have yielded so easily to copper barons such as the notorious Marcus Daly. A veteran campaigner out of the no-man's-land of Irish poverty who had fared well and ill over the battle-fields around Virginia City, he had been engaged by the Walker brothers of Salt Lake City to purchase a silver mine for them in Montana, the Alice, which cost them 25,000 dollars, over which he became superintendent with several holdings of his own. His meteoric hurtling to fame in the Haggin, Hearst and Tevis syndicate for the purpose of developing Michael Hickey's Anaconda mine is a legend in mining history. One of the most powerful teams of mining experience and business acumen in the West, it raised Daly to Olympian heights on the mounds of silver, gold and copper that stretched from Alaska to the furthermost tip of South America[1]; and in Montana he fought tooth and nail for possession of the very heart of Butte. Daly made an alliance with Cornelius Kelley and "Dan" Hennessey, one of the most astute business men of Butte, to control the city and Montana for the Democrats. His rival was William Andrews Clark, the leading banker and merchant in Montana who, at the age of thirty-three and already a millionaire, was equally determined that Butte and Montana should be saved for the Republicans[2]. The upshot was that Daly imported Irish labour to replace the Cornish in his Anaconda and Never Sweat mines; and that the episcopalian Clark, in alliance with that other phantom of Butte, Frederick Heinze (who arrived there in 1889 penniless and absconded in 1906 with almost 10,000,000 dollars which he thoughtlessly lost through "playing" the New York Stock Exchange) fought back on the Republican votes of the Cousin Jacks.

The natural antagonisms of Catholics, Episcopalians, Methodists, Democrats, Republicans, Irish and Cornish were raised to flash point and open warfare by these men, especially at election time when the rival candidates would be introduced into the boxing and wrestling rings, but alas not to fight: that was left to the spectators. The mining camps of course were accustomed to this display of fisticuffs by Irish and Cornish, but the sinister feature about Butte was that the warfare was systematically waged underground, the Cornish being almost drafted into the "maquis" of Heinze who ordered them from the bottom of his Minnie Healey, Cobra, Rarus and McDermit to infiltrate the workings of the rival syndicate, causing Daly to spend almost 1,000,000 dollars in law-suits brought against him over the obscure Apex Law[3]. They ran high-pressure hoses into enemy territory, blew powdered

---

[1] Isaac F. Marcosson, *Anaconda* (New York: 1957), p. 35.
[2] Ibid., p. 84.
[3] Ibid., p. 112. Marcosson explains the apex law thus: "In the absence of mining law, the first miners in California, to protect their rights, had resorted to Spanish law about ownership and Cornish law about claims, i.e. a miner should possess a vein in its dip into the earth between the vertical boundaries of the claim. Friction

lime through one-inch pipes, padlocked bulkheads and even manufactured hand grenades by packing empty tomato cans with dynamite[1]. For almost a year the boarding houses of Mrs. Burke and Mrs. Rowe, only a few yards from each other, hummed with charges and countercharges about dynamite with split fuses being sent down the shafts in buckets, and arguments about the powerlessness of the Butte Working Men's Union to end this lawlessness that might in time destroy them. It was only stopped when Rockefeller bought out Heinze and then broke him on the rack of Wall Street, thus sparking off the financial panic of 1907. Mr. Herbert Hamlin remembers the strain of those days; he never dared to go on the graveyard shift at the Pennsylvania without a gun; and he sent his wife back to New York when he found three sticks of dynamite in their ash-can.

Much as he admired the Cornish, he is critical of their willingness to opt out of their political responsibilities, since it made them a sitting target for Heinze and then the gullible victims of agitators from the left-wing organization, the Industrial Workers of the World. The picture perhaps is neither complete nor entirely true, for there was a powerful moderate opinion among the Cornish that sought refuge in an independent press. Bored and frustrated by *The Anaconda Standard*, owned by the Democrat Daly, and by Clark's Republican *The Butte Miner*, T. H. Dunstan and R. J. Oates in 1894 bought out *The Populist Tribune*. While the aim of the paper was said to be "for the home as well as for the business man", it is significant that "Populist" was discreetly dropped to remove any possible doubt about its non-political affiliations. *The Tribune-Review*, as it was called in 1898 until its disappearance under the depression of 1920, was to be mainly devoted to "News from Cornwall", culled from the columns of *The West Briton* and *The Cornishman*, whose editor, Herbert Thomas, had roved over the mining West.

The Tribune, politically self-purged, bound the Butte Cousin Jacks to "home" as they worked the properties of W. A. Clark down in the

---

arose because of overlapping veins and in 1872 Congress passed what became known as the Apex Law. The intersection of a vein at the surface is known as the apex and, according to this law, the owner of a vein was entitled to follow that vein downward even though it might lead under the surface holdings of other claims. Thus a miner might find himself working far beyond the area of his original claims. This would not have mattered if the veins were continuous but more often than not they were crooked or faulted. The problem in Butte arose from a gap in the law which Heinze exploited. If a vein leading from the surface was lost near the vertical side-wall of a claim, and a similar vein was found below it or to one side in an adjoining claim, who was to decide whether the second discovery was a geological continuation of the first?".
[1] *The Montana Standard-Post*, 24 November 1963.

Travona, Mountain View, Tramway, Colusa-Parrott, Original, Moulton, Stuart, Black Rock, Poser and the Speculator, where Captain Bill Webb, the assistant superintendent, has been remembered as "a real miner, raised in Cornwall but never learned to speak English". William Owsley, Lee Mantle, W. H. Davey and William G. Cocking were all mayors of Butte[1]. In 1895 Sam Williams produced *The Pirates of Penzance* with a cast that was entirely Cornish, probably drawn from members of the Butte Carol Club that had been founded by Alfred Paynter about 1880. Cornish mining preachers lay thick on the ground in direct proportion to the work of reclamation on "the glittering hill" but there never seemed to be any shortage of labourers busy with their sickles of personal salvation. For instance, we hear of the young Popes who immediately after their marriage at St. Keverne in 1891, set out for Montana to establish Congregational churches throughout the Yellowstone Valley; and of Samuel J. Hocking, six years pastor at the silver camp of Neihart who, when it was destroyed by the repeal of the Sherman Silver Purchase Act, refused to accept another living at twice the salary because he felt "that the salvation of his congregation was of more importance to him than ease and support"[2]. And at Ennis, surely a Cornish place-name, Mrs. Hetty Bennets, now over eighty and almost blind, recalls that her father from Camborne died when she was five; that her mother was driven to earn her living by taking in boarders until she married another miner who died of consumption in South Africa; and that she herself as a girl of nine had to mother her half-brothers and sisters so that she was deprived of all schooling.

William Pearce was one of the most lonely rangers in the mountains and plains of the Montana that is "high, wide and handsome". Born at the fishing village of Newlyn he was bound apprentice to a blacksmith in the nearby hamlet of Sheffield. Finding little work locally and being ambitious, he thought he could better himself in the dismal and squalid port of Cardiff, but his wages amounted to only 4 dollars a week. So about 1902 he went off to Canada on the £8 he had saved, arrived almost penniless on its west coast and was surprised to find that the only credentials for work were ability and thoroughness. So he wandered through Washington, Oregon, Montana and Idaho, mostly along the railroad tracks, carrying the tools of his trade on his back. Then, as was the habit of the Cornish, he returned to Newlyn for a wife and set her up in a house that he built by the side of his blacksmith's shop in

1   *The Montana Standard*, 20 April 1958.
2   E. L. Mills, *Peaks and Pioneers* (Portland: 1947), pp. 114-5, 201.

Seattle. But she never liked the place and returned home, while he trekked once more along the "main-travelled" roads until he pitched down at Grass Range, a railroad town already in the first stages of its decay. Here he heard that his wife had died. He returned to Cornwall and married again, but pined the remainder of his life away, a frontier American in an alien England that seemed to turn to socialism as the panacea for all its ills. As he once wrote in *The Grass Range Review:* "Socialism will not flourish in a country where the mayor's and cobbler's sons sit side by side in school, where the bankers and the dustmen address one another as Tom and Jack. America has achieved much in her one hundred and fifty years of nationhood. May she climb to greater heights, framing her policy in accordance with those famous words, 'Government for the People and by the People', immortalised by that grand old woodchopper, Abe Lincoln".

Charles Bishop's mining maxim derives from the same experience: "A shift's work plus a little extra for a shift's wages makes you independent and free to demand your just due". And he should know for the story of his family is one of continuing neglect and hardship. His grandmother was only four when her father left for America at the end of the Civil War; she brushed the tin dust from window sills and filled the heavy wheelbarrows with ore. When she was twelve, her father wrote to her mother that he had taken a "common-law" wife who was now dead, that she had given him five children, and that he wondered whether his Cornish wife could now join him. Since the passage money was with the letter and "there was nothing for women in Cornwall", she agreed and went. But Bishop's grandmother struck out for herself and managed a boarding house in Calumet, Michigan, where she married a Redruth miner, John Murton and, with seven children, embarked on a life in the wilderness of mining camps, sharing their troubles. At Tonopah in Nevada one son died from drinking the water in a horse trough, a second was decapitated in the Silver Queen and a third dropped dead from heart failure. These calamities so unnerved one of the three daughters that she determined to break with mining and studied business management in San Francisco. But she fell under the compelling fascination of Nevada and married an English miner, Cleveland Dow Bishop.

These were the parents of Charles Bishop who, when only a few weeks old, was taken to Butte where his father mined for a time and then died, comparatively young, of heart failure; and in a tent in the back-yard the old John Murton coughed his life away through his sick lungs. For his grandson, Charles Bishop, there seemed no escape from a similar fate unless he chose another occupation.

It was in the depression of the early 1830's that he entered the mine, mucking rock into 1600 lb cars and tramming them out of the workings at the rate of twenty cars a day compared with the thirty-five that the shift boss required. Then he found himself at the 3200 ft level in the "Chinese laundry" workings of the Steward mine, where the temperature was 120°F and the humidity 86%: "This I could take but not the bad air, for my working days were now one of wading through copper water halfway to my knees, tramming car after car of rock with a head and stomach full of gas". He was possessed of a fear of working "in a strange place", and witnessed his first accident when "a stranger" to the stope fell sixty feet to his death. He was forever tormented by the real possibility of being crushed to death by fast moving cars or being impaled on leaden pipes. And when he had trained as a hoist engineer he was startled on hearing for the first time the angelus of nine bells that signalled the flight of some poor soul, and in a sweat of terror watched the body brought into his engine room to await the arrival of the "funeral car". So "deep enough" it was, and Bishop exchanged the mine for the engineering of refrigeration.

Today, the Cornishman in Butte, as elsewhere, has been lost in the crowd; not a recent phenomenon—elegies about the death of the past were being sung and intoned fifty years ago. And as the Rev. George Sloan, the builder of Presbyterian churches from Arizona to Montana, says: "The Cornish in Butte seem rather to want little to be said or written about their virtues or success, and those I have known best seem to desire only that they may be worthy of respect"[1]. Even Joe James from Townsend, near Hayle, was reluctant to describe how he was one of twenty-five to be miraculously rescued from the Granite Mountain shaft in 1917 after being entombed for forty-eight hours, while 200 of his companions died in one of the most disastrous underground fires ever recorded.

Such were the men who carried their calculated risks from Cornwall to Butte. This spur of the Continental Divide was a highly mineralized area four miles east to west by three miles north to south. Together they sank a prodigious total of forty miles of vertical shafts, drove over 2,000 miles of underground passages, blasted out more than 6,000 miles of stopes and made the Anaconda Company the world's greatest producer of copper, silver, gold, zinc, lead, manganese, aluminium and arsenic, with mammoth interests in Nevada, Connecticut,

[1] See an article on Captain Tommy Couch in *The Roy Enterprise*, 7 January 1918.

Idaho, California, New Jersey, Indiana, Wisconsin, Illinois, New Mexico and, last but not least, Arizona.

One of the youngest states of the Union, Arizona was still a Territory when the last of the Cousin Jacks were leaving Cornwall at the end of the century. But it was copper country quite different from any other they had worked in. The first miners had parted its long grasses on their way to California in 1849; the first Europeans had come that way from Spain in 1540 under the leadership of Coronado to find the mythical Seven Cities of Cibola; and it harboured a significant Indian population that was deadly in its hatred of the whites. Today citrus, cotton, cattle and climate are as important as copper; and the State seal understandably depicts a range of mountains against a blazing golden sun, a storage reservoir that irrigates orchards and range, and a quartz mill before which leans a miner with pick and shovel. But in the early days immediately after the end of the Civil War it was cattle that was king, and lawlessness and inaccessibility the chief deterrents to settlement.

In 1874 the first dazzling reports of thousands of shallow deposits of gold and silver began to attract prospectors, but transportation proved to be their first problem. For those from the eastern states, the best and shortest way was through Albuquerque along a trail that provided usually a fair abundance of grass, water and wood; but, even so, the distance from St. Louis was almost 800 miles. For those emigrating from the Mother Lode there were three possible routes: a 400-mile haul by road from San Bernadino to Prescott; an equally rugged crossing of the desert from San Diego to Yuma and on to Tucson by stage that cost 90 dollars; or by steamer from San Francisco once every three weeks up the Gulf of California, and then by river steamer to Yuma. For many years, much of the freight for the mines came this way through Yuma, one of the hottest spots anywhere in America; most miners had to pass through its torrid heat after the exhausting perils of the Southern Colorado Desert where, even as late as 1890, so many prospectors had died of thirst that iron posts, spaced a mile apart on the main compass bearings, had to be erected to guide them to water-holes[1]. From Yuma to Phoenix the desert was equally relentless, many a traveller being lost in the drifting sand until it was partially tamed by the laying of a plank road.

In those early days Phoenix was hardly more than a stinking hole for drunks. Today, rising on cushions of hot air by day, it seems, and suspended like a necklace from the night sky, one would never guess

---

[1] *The Arizona Silver Belt*, 12 April 1890.

that its "founder" was the morphine addict, John Swilling. Though he died in the notorious Yuma jail for the only crime he did not commit (holding up a stage at Wickenburg), in a rare moment of inspiration that seeped from his fuddled brain he recognised the aptness of the name that his partner, Darrel Duppa, suggested for his settlement: Phoenix, a burst of beautiful sunshine[1]. Miners moving south to Tucson, however, had no time to meditate on the imagery, for before them stretched more desert. Even today this route, from the safety of a railroad train, both attracts and repels. Cattle shelter in immense stockyards under corrugated iron roofs to escape from the intense heat; a church and a cemetery suddenly appear in the middle of nowhere, the crosses bleached as white as the sand they stand in; and between the Sacaton Mountains to the south and the Superstition Mountains to the north dirt roads cling almost desperately to the railroad track. Miners voyaging further east would lose themselves in a proliferation of nature that was gestated, it seems, in a succession of nightmares. Towards Picacho Peak and Red Rock looms the alarming Saguaro Forest with its giant cacti of every conceivable shape and kind: the saguaro that climbs to a height of forty feet and can weigh ten tons; the cholla or "jumping" cactus; the ocotillo or "cane" cactus, and the prickly pear and the passajo.

Tucson was no place for Cornishmen who went down to the sea in ships or mined beneath its waters, for here the rains came only for a few days in the year and the wind was forever scorching. Its past had always reeked of fire and lightning; as the winter rendezvous of the mountain men, the hunting geographers of the West and terrifying in their cups; and as the hide-out of deserters, renegades and cattle-thieves in the turbulent days that followed the defeat of the South. The older parts of this frontier town still show signs of that nerve-stretched life in a huddle of ramshackle adobe huts that are the last refuge of shiftless Mexicans, despised Negroes and dispossessed Yaqui Indians. The graceful town houses in Snob Hollow, two in the English half-timbered Tudor style, were built by cattle barons who brought some semblance of order and civilization; but the dreary waterless waste of dirt tracks, mesquite and tumbleweed beyond, is relieved only by the glistening white walls of the mission of St. Francis Xavier Del Bac. It is all an ironic comment on the present lot of the Papago Indians who gave the name to this State when it meant more than it does now, for Arizonacs is their word for "a small but ever-full stream".

---

[1] Andrew Wallace, 'John W. Swilling', *Arizoniana* (Spring 1961).

It was Bisbee rather than Tucson that attracted some of the Cousin Jacks to Arizona. It came to life in a canyon of hard grey limestone in the shadow of the bright red earth of Chihuahua Hill and the bronzed summit of Sacramento Hill which both covered rich green malachite, blue azurite and pure native copper. Among the 500 or so claims none was richer than the Copper Queen, set on a hill that hoarded an almost circular body of oxidized copper, iron and manganese, about 60 ft in diameter and enclosed in unaltered limestone. The modern excavation of these cores has been the work of geologist and petrologist, but in the beginning they were sniffed out by the old-fashioned miner, whose occult, mysterious and intuitive sense led him to anticipate the ore he could not see[1]. Since the first method of reaching the ores was by shafting and square-cut timbering, it seems fairly certain that this must have been partly the work of Cousin Jacks. Indeed the Copper Queen Branch of the Phelps Dodge Corporation has records of some of them.

William Hugh from Grampound and Michigan was in 1906 employed as a miner at the Holbrook and Czar divisions. John Ferrel had a varied career as a hoist engineer at the Chapin mine at Iron Mountain in Michigan, as a power plant engineer in Colorado and as the engineer in the smelter at Globe, before settling down to mine for the last ten years of his life in 1906 at the Copper Queen. James Hart was four when his parents emigrated to Virginia City in 1875; there he trained as a blacksmith, married a Cornish wife, moved to Redding in California and Cottage Grove in Oregon, finally finishing out his days at the Copper Queen in 1907. Peter Andrews was also four when his parents took off in 1877 for Nevadaville in Colorado but, when his mother died, his father sent him back to Cornwall to stay with relatives until he was fourteen, when he rejoined his father in Silver City, Idaho. Bisbee finally claimed him as their shift-boss for twenty-two years where he worked alongside James Cowling[2]. The reasons for staying so long in one of the world's most outlandish places remain their secret. Most workers in 1902 found the Bisbee mines too hot at a depth of 4,000 ft; water was hawked by Mexicans at the price of half a dollar a sack and was too alkaline to drink; and there were only two business houses that amounted to anything; the Copper Queen Merchandise Store and the Copper Queen Hotel, both owned and operated by the mining company. Otherwise it was a town of gambling saloons, fast

1 James Douglas, 'The Early History of Bisbee District', *The Bisbee Daily Review* (3 August 1931).
2 *The Copper Queen Bulletin*, June, December 1923; March 1928; and March, May 1930.

women, Mexicans and Yaqui Indians; so most miners worked only long enough to qualify for a pay-day and then drifted off, like William Vine of Central City.

If they settled at all, the Cousin Jacks generally preferred Globe and Miami in the cooler uplands of central Arizona. Globe displayed one striking feature that marked it off from all other mining camps: the presence of the Apaches. Their San Carlos Reservation, which was established in 1872, consists of 1,500,000 acres of open range, where today 5,000 Apaches work the most extensive cattle ranch in the world. Ever since then, and long before, this area of Arizona has been bloodied by the fight between miners, cattle-rustlers and Apaches for possession of land. At the top of dark masses of basalt and crenelated granite looms Apache Leap, from whence they hurled themselves to their death rather than surrender to the American cavalry. The cemetery at Globe, white against a bare hillside, its only shade coming from the cypresses planted by Cornish miners, tells the tale of violence and robbery in which white fought against white and both against the Apaches. Here are buried the Pascoes, the Opies and the Trevillians of Cornwall, their names cut in plates of copper screwed to slabs of granite that strangely compare with the memorials of others: brown concrete pillars that resemble the trunks of trees dug out of some petrified forest. Nearby are the graves of Sheriff Glenn Reynolds and his deputy, "Hunkydory" Holmes, murdered in October 1889 at Riverside by the Apache Kid whom they were escorting to the state prison at Yuma[1]; of the Indian scout, Al Sieber, for the wounding of whom the Apache Kid had been arrested; of the Apache Nah-deiz-az who was hanged publicly in Globe for the murder of an officer stationed on the San Carlos Reservation; and of Phin Clanton, cattle-rustler, desperado and brother of another outlaw, gunned down by Wyatt Earp. And all these headstones of miners and renegades look out toward Pioneer Pass through the Pinal Mountains, the trail used by the troops; and across the canyon to Pascoe Hill, sometimes called Pasty Hill; and to Cousin Jack Hill, sometimes called St. Just Hill.

In Globe today other Cornish place-names are still remembered. A clerk in the Court House claims that her grandfather, Albert Tallon, emigrated from Penzance to Michigan, then to Idaho and finally to Globe, where he became the safety inspector at the Old Dominion mine. Her colleague reveals that her father, Albert Parkin, born in Wadebridge to a stonemason who dressed the granite blocks for the Eddystone lighthouse off Plymouth, was taken as a child to New

1 See Jess G. Hayes, *Apache Vengeance* (University of New Mexico: 1954).

Jersey, grew up in Michigan's cold Calumet, and then mined in Prescott before settling his family in Globe. Art Trevillian recalls that he once worked at the bottom of Dolcoath mine at Redruth. Mrs. Knight, a frail old lady but as bright as a meadow-lark, recounts how she went alone from Grampound to Globe to join her husband and then together they journeyed "to sink a shaft in Mexico".

These are the old-timers whose occupations have gone, not with the onset of old age, but with the closing of the one mine on which depended for so long the prosperity of Globe: the Old Dominion, an old-timer too with its timbered shafts, its skips and its hand-drilling. Today, as in Montana and Utah, all the mining is of the open-pit method, vast amphitheatres of engineering precision where the greens and the pinks of the copper seams are exposed in all their natural beauty to the blue of the sky. For this is the only economic way to exploit the poorer ores which yield only .4% of copper; richer ores of .7% copper alone would justify the costlier process of underground mining. One open pit, which had been cut in 1954 after the surveyors had marked out the exact datum points on the mountains where the first blasting would begin, reveals a massive operation of working surfaces called "benches", from which, by a continued excavation, a series of tiers or terraces leads down to the core of the pit, where the drama of dynamiting and clearing is enacted by mechanics, not miners, who scrape the mountain down to its bare bones with their monster mechanical shovels. The crushing plant that has taken the place of the old stamp mills are miracles of power; gigantic bins and drums crush the rock to gravel and dust before the flotation process takes over, in which pine oil recovers the copper and then passes it on to the smelter. An older pit that appeared to have finished its days is "cleaned" by washing the benches with sulphuric acid, the residual copper sulphate being collected from the tunnels and drifts at the 600 ft level and then pumped to the surface, to be run over ordinary household tin cans on which the copper is finally collected.

It is not likely that there were any Cousin Jacks in Arizona immediately after the Civil War, for it was far too remote a country and few of them were lone wolves. Even in 1870 there were only 1039 settlers engaged in mining out of a total population of nearly 10,000. Yet it seems that the Cornishman was already the symbol of miners generally: "The case of Richard Roe as, with swinging pick, he follows his tiny vein, also has its interest (for) he was wondering just what he would do with the million dollars he would get . . ."[1] Later the imaginary

1 Will Robinson, *The Story of Arizona* (Phoenix: 1919). p. 250.

figure of Roe gives way to the legendary John Pierce who, however, was real enough. In 1894 he was living in Cochise, not far from Tombstone, where his wife ran a boarding house. Deciding to quit mining, he bought a herd of cattle on the range in Sulphur Springs Valley and sent his feckless son to Kansas City to sell them, but he gambled away the profits. Musing over the iniquities of the world with another Cornishman, John Kinsman of Globe, John Pierce picked up a rock, felt it and took it to Tombstone to have it assayed; as he expected, it proved to be a bonanza, assaying at the rate of 80 dollars of gold and 20 of silver to the ton. In true Cornish fashion, the claim was shared out among the family in equal parts but, since they had no capital, was sold on the market to Richard Penrose of Cripple Creek in Colorado for nearly 300,000 dollars. Mrs. Pierce apparently drove a hard bargain, refusing a bank draft for the first 50,000 dollars, demanding cash, and reserving for herself the right to operate the only boarding-house as soon as the mining operations began. At its peak the mine yielded as much as 300,000 dollars' worth of silver a month and, during the twenty-four years of its life from 1894 to 1918, produced 15,000,000 dollars' worth of gold. Today the memorial to this extraordinary transaction is the ghost town of Pierce[1].

Some Cousin Jacks, of course, would work in the most inaccessible places, like those who dug out the King of Arizona gold and silver mine in the Kofa Mountains in 1898, way out on the desert some fifty miles north of Mohawk on the Gila River. Water was the major problem here, for it had to be transported in whisky barrels by mule teams from Mohawk, a journey that took three days; after being swilled around in the heat for so long, it made the miners sick. Yet within a year Kofa somehow or other was supporting a small Cornish community. As one miner has recalled: "The foreman was a heavy-set Cornishman. I forgot what we called him—but we called him anything but a good name—and after he got in, most all the help was Cornish. He had Cousin Jacks coming in all the time"[2]. Water was not the only problem; equally deadly were the sulphur fumes and several of the Cornish never recovered. But, in spite of these hazards, it seems that there were Cornish women at the camp, according to another American miner who knew them and who has described their arrival in the care of the local stage-coach driver glorying in the name of Macbeth: "If he

---

1 See Richard E. Sloan, *Memoirs of an Arizona Judge* (Stanford: 1932), p. 99; and Nell Murbarger, 'Ghost Towns of Arizona', *Arizona Highways* (August 1960).
2 E. B. Hart, 'First Days at the King', *The Calico Print* (May 1953).

came walking the team, saving it, you knew he had on some fresh
people from Cornwall and that he was going to give them a thrill . . .
and he'd pull the four horses back on their haunches and you'd hear
more squealing and the whole bunch of passengers would be a pile in
the front of the stage"[1].

Thus gold was king and silver was queen long before copper became
the mainstay of the Arizona economy; and in their early exploitation
the Cornish had little part to play. And this is understandable for,
when Charles D. Poston, "the father of Arizona", was developing the
Heintzelan mine near Tubac, first opened by the German Herman
Ehrenberg in 1854, they were making the overland journey from Wisconsin to California. To them Arizona was only a name. They knew
nothing of the Arizona Mining and Trading Company that, incredible
as it may seem, was shipping its ores from Ajo to Swansea in South
Wales for smelting; nor of the galaxy of golden mines that burst
around Prescott; nor of the discoveries of William Wickenburg
who in 1863 found gold so near the surface that he was able to pick it
out with his knife. Near Wickenburg was located the Vulture mine
where the superintendent was the Cornishman, Cyrus Gribble; on the
night of 19 March 1888 he and his partner were escorting a load of
bullion to Phoenix when they were murdered by a Mexican[2]. Incidents
like this were often reported in the Californian newspapers and no
doubt deterred many a young Cousin Jack who contemplated trying his
luck in Arizona, where rich hauls of silver were being taken out of
Pinal County and where by 1876 no less than 975 claims had been
filed[3]. In Pinal the most famous were the Silver Queen, first located in
1871, relocated in 1875 and today the main shaft of the Magma copper
mine at Superior; and its peer, the Silver King, first discovered in 1871
by a soldier of the name of Sullivan.

It was at this time, some ten years after Congress in 1862 had created
Arizona Territory, that newspaper editors began a drive to induce
miners and capitalists to go to Arizona. On 1 April 1872 *The Weekly
Arizona Miner* asserted that it was "the duty of every Arizonian not
engaged in quartz-mining to encourage and help all who are". On
18 April the editor launched an even more impassioned appeal: "Many
of the mines already spoken of are lying idle, waiting for capital to

1 William G. Keiser, 'Cornishmen and Chinese Cooks', *The Calico Print* (May
   1953).
2 *The Arizona Silver Belt*, 24 March 1888.
3 Robinson, p. 254-7.

work them; thousands of stock ranges await men who wish to engage in this business; tens of thousands of noble pines await the axe and the mill". Yet he knew that he carried his life in his hands. The founder of *The Weekly Arizona Miner*, Richard C. McCormick, and his first editor both worked in 1864 with rifles strapped to their backs and the second editor was shot by Apaches in his office[1]. What Arizona needed, it was said, was a large immigration of sturdy and hard-working miners with their families from Britain and Europe where "servant girls in private families often get but ten dollars a year". Decent wages at Prescott, three newspapers, the electric telegraph and the coming of a railroad were held out as additional inducements; and labourers could earn three dollars a day, blacksmiths and miners five, and carpenters eight.

But as yet the Cornish in Nevada City and Grass Valley were deterred from venturing into Arizona by stories of Gila monsters, the fierce heat and the raids of the Apaches. In 1871, the year that Ben Reagen found on the western slopes of the Pinal Mountains the 300 lb ball of almost solid silver, shaped like a globe with surface scars that resembled the continents of the earth and that gave the name of Globe to the camp just inside the San Carlos reservation, some forty citizens from Tucson together with more than 100 Papago Indians slaughtered eighty-five Apaches, all but eight of them women and children. Globe seemed no place for settlement until the army moved in and until enough claims had been filed to warrant the forming of a mining district around the Silver King bonanza of 1875. But even then, Globe was forbiddingly isolated; all supplies had to be brought in on pack animals, while the nearest place for smelting was at Silver City in New Mexico. Not until after 1882 did the Concord stages cross Pioneer Pass over Chalk Hill and then down to the Gila River, where the steep hills and the sandy washes often required passengers to get out and walk; until then immigrants had to be content with a buck-board trail to Pioneer mining camp[2].

We know of three Cornish families who worked at the Silver King: Richard Trevathan, George Lobb who married his daughter, and the Knights from Penzance. John Knight brought an entire family of nine from Grass Valley; he was a timberman at the Silver King but, with commendable foresight, also opened a store and, when the King failed

1 Joseph Miller, *The Arizona Story* (New York: 1952), p. XV.
2 See articles in *The Globe Recorder* (February and March 1956) by Mrs. Clara Woody; and Casper R. Smith, *The Arizona Silver Belt Anniversary Edition* (8 May 1958).

to survive the devaluation of silver, simply moved his business to Tempe and survived himself. Richard Trevathan might have been one of the first of the Cornish in Globe for he appears to have been occupying one of the seventy-four adobe houses that Judge Aaron H. Hackney found when he started publishing *The Silver Belt* in 1878. On his arrival a letter was reported to be waiting for him at the office of the postmaster, Edwin M. Pearce, who had to collect the twenty-five cents charged by the Indian runner who had carried it from San Carlos thirty miles away[1]. Where the Trevathans had come from is not known, though it might have been Leadville in Colorado, for arrivals in Globe at this time were complaining of the altitude there[2]. Perhaps they included among their number Jim Jewell who was reported as having run a tunnel into the Little Raven lode at Gold Hill[3]; John Blewett who was leasing part of the Rescue mine; another Jewell who lived up to his name by producing from his Lost Gulch mine "some beautiful specimens of ore literally filled with pure gold"[4]; and John Trewartha, who had made his way from Wisconsin and boarded at Pascoe House in Globe.

There is not much doubt about the Cornish origins of the Pascoes. *The Great Register* of 1881 lists four brothers: Benjamin, James, James H. and Thomas, who were born in Galena, Wisconsin, or in California; and a fifth, Benjamin F., born "in England". Apparently the latter was the first to arrive in Globe, he being the eldest, and he very quickly achieved a successful combination of mining and business. He not only owned the Petaluma mine (a Cornish name) but set himself up as an hotelier to catch the trade of the incoming miners[5]. His brother Thomas followed him to Globe in 1881, establishing the Pascoe Livery Barn and a hay and grain business to go with it at the back of McNelly's saloon. Moreover, he became an influential member of the Globe Stock Growers' Association[6], and was a government forage agent, there being "no one better qualified to meet the requirements of the contract"[7]. He was Vice-President and Director of the Gila Valley Bank and Trust Company, and Eminent Commander of the Globe Commandery of the Knights Templar; but, not surprisingly, for this is a certain Cornish trait, "he never entertained political aspirations"[8].

1   *The Arizona Silver Belt*, 17 October 1878.
2   Ibid., 6 December 1878.
3   Ibid., 7 February 1879.
4   Ibid., 2 May 1879.
5   Ibid., 24 January 1879.
6   Ibid., 7 January 1888.
7   Ibid., 29 June 1889.
8   *Who's Who in Arizona* (1913).

His brother Benjamin F. owned Ash Springs Ranch, where he fed and watered eighteen horses and some 200 head of cattle, and was Sheriff of Gila County from 1883 to 1886[1].

By 1879 communications with Globe were becoming easier. The Southern Pacific was pushing its rails across the desert east of Yuma at the rate of two miles a day while, to the north, the Santa Fé had topped the Eaton Mountains, though goods destined for Tucson still had to be hauled by ox-trains across the plains of New Mexico[2]. Judge Hackney also broke down the isolation of Globe by linking the office of *The Arizona Silver Belt* by telegraph with every military post in Arizona, while in his editorials he persistently urged miners to come south, and especially those in Leadville "where people die by hundreds of diseases peculiar to the climate and are buried at night to avoid alarm"[3]. Claiming that his *Silver Belt* was "a paper for the miner, a paper for the farmer, a paper for the mechanic, and a paper for everybody", on 27 August 1879, when the thermometer stood at 104°F, he wrote ecstatically of the attractions of the new gold and silver camp: "Globe as a place of residence has many things to recommend it . . . the climate is exceptionally good and the town is nicely laid out . . . the summers are not exceptionally hot, nor the winters extremely cold".

Hackney carefully avoided any mention of the Apaches and played down the rigours of riding on the back of a mule in the dust of the noonday heat; so in 1880 Globe was booming, its busy canyon alive with the thunder of sixty-six stamps, pounding away at the ore from a cluster of mines. From the Golden Eagle, the Andy Campbell, the Golden Nugget, the Moffat and the Eureka came the gold; from the Stonewall Jackson, the Mack Morris, the Independence, the Alice and, above all, the Old Dominion, came the silver; and from the True Blue, the Carrie, the Tacoma and the Illinois the first copper. The Cornishman George Millet is reported as owning 200 shares in the Moonlight Mining Company[4]. Trewens, Moyles and Hockings donate money for the building of St. Paul's Methodist Church[5] and the foundation stones are laid in April 1880. For another seven months the services and the Sunday School continue to be held in the offices of *The Arizona Silver Belt;* and then in November it is dedicated, Tom Pascoe having given both land and money for the building of a parsonage. Hackney's articles

1 *The Arizona Silver Belt,* 21 January 1888. Pascoe committed suicide shortly after relinquishing office.
2 Ibid., 3 June 1879.
3 Ibid., 20 June 1879.
4 Ibid., 3 January 1880.
5 Ibid., 14 February 1880.

were reaping their harvest: "Globe is rapidly filling up with strangers, many of whom are here with a view to investing in mines and others for the purpose of prospecting to cheer their loved ones at home"[1]. But not all strangers were welcome, for one miner was complaining that Globe was too full of schemers, black-mailers and claim-jumpers[2]. Many more Cornishmen no doubt appeared in 1881 when the Southern Pacific made contact with the Santa Fé at Deming in New Mexico, for this is the year when the *Great Register* yields its first significant crop.

It mentions James P. Faull, Harrison Jewell from New Hampshire, Thomas Jennings, Philip Oates and William A. Trerew; Elijah Trengrove, naturalised in Santa Clara County and therefore almost certainly a miner from New Almaden; Richard Trevarthan who fancied himself as a humorist[3]; R. H. Trevathan who married a Mexican, Mary Gonzalet, in a Globe saloon[4]; William Trevillian from the Mother Lode; Richard Trewartha; and the five Pascoes. By the following year the new arrivals were: James Argall, Thomas Curnow, William James and Harry Hancock, all from Michigan; Henry Chirgwin; William Eddy from Reno; Joshua Nicholls from Montana; Thomas Nance, Arthur Mitchell, Henry and William Paul from Central City (the latter being in charge of the La Platte mine near the Mack Morris)[5]; and twenty-five others whose Cornish background cannot be determined with any certainty. Globe never seems to have been a completely Cornish town but here, as elsewhere, the Cousin Jacks left their marks on the social and religious life of the community; for instance, H. Faull became one of the trustees of the Methodist Church and the superintendent of its Sunday School for "he has always been a devoted friend of the children"[6]. But all too soon the silver gave signs of pinching out, and the town would have been doomed but for its copper.

The first copper to be mined was on the Hoosier claim of the Phelps Dodge Company, James P. Faull being given the honour of driving in the first pick on Thanksgiving Day 1880; he was a Cousin Jack of some standing, for he owned fourteen claims himself. But it was the almost forgotten Old Dominion claim that was the true saviour of Globe. According to *The Phoenix Gazette* of 5 April 1883, its shipment of copper for one week from its two 30-ton smelters amounted to 254,000 lbs; from 1882 to 1885, at the same rate, this amounted to a

1 Ibid., 3 April 1880.
2 Ibid., 10 April 1880.
3 Ibid., 16 January 1889.
4 *Magistrates' Reports*, County Court House, Globe.
5 *The Arizona Silver Belt*, 29 April 1882.
6 Ibid., 9 June 1888.

grand total of 23,000,000 lbs. This was a tremendous achievement, for the coke for the furnaces, which burnt 100 tons a month, had to be shipped all the way from Cardiff in South Wales round Cape Horn to San Francisco, then by railroad freight cars to Willcox, and finally hauled the last 120 miles to Globe in wagons pulled by 14-mule teams; this last part of the journey took two weeks if the weather was favourable, and much longer if the Gila River was flooded. *The Arizona Silver Belt* now predicted that the time would come when Arizona would produce copper more cheaply than Lake Superior[1], though this seemed premature in 1887 when the plant had to shut down because of a fall in the price of copper, an awkward time for some Cousin Jacks who had recently arrived like Ben Faull, George Lobb from Deadwood in South Dakota, Joseph Kinsman, Thomas Wakefield from Penzance and John Trennary[2]. However in 1888 it was reopened and a new shaft, the Interloper, sunk in an effort to clear the bank overdraft; the venture produced dividends for the output of copper in 1888 rose to 4,640,000 lbs out of a total of 17,000,000 for the whole of Arizona[3]. The centre of gravity for copper was unmistakably shifting from Upper Michigan, sending its Cornish miners scurrying either to Montana or to Arizona.

In many ways, Globe was far more attractive than Butte. Sited in a long narrow canyon surrounded by foothills that reached to the Sierra Apache and the Hayes Mountains, it had a mean temperature of 63°F and an average rainfall of only 15.72 ins, though this could often be exceeded; the Pinal River in 1891 burst its banks, destroyed T. A. Pascoe's corral and house at Wheatfield and almost swept away that of Elijah Trengrove[4]. Cousin Jennies no doubt thought twice about the snakes and tarantulas, but at least Globe was served by a Cornish doctor, Willie King from Redruth. The main problem, however, was still the lack of a railroad. If they came from Michigan on the Santa Fé from Chicago they still had to finish the last 123 miles from Willcox by stage; if they came from California, they could only get as far as Casa Grande by Southern Pacific, and the last ninety miles had to be made also by stage by way of Florence, Silver King and Devil's Canyon. Not until December 1898 did the first train steam into Globe.

Yet we find Carrie Oates making one of these tiring journeys and settling down to become one of the stalwarts of the Methodist

1 Ibid., 7 January 1877; 22 May 1882; 14 February 1885.
2 *The Great Register of 1886.*
3 *The Arizona Silver Belt*, 15 January 1889.
4 Ibid., 21 February 1891.

Church, entertaining its members with her music when they were not being subjected to lectures on such themes as "Corrupt literature" and "Jordan is a hard road to travel"[1]. From Cornwall too in June 1888 arrived the wife, daughter and son of Richard Trevarthan, after a separation of fifteen years. What they thought of this rough frontier town can only be guessed. The feud between the cattlemen of the Graham and Tewksbury families was at its bloodiest. Thomas Pascoe saw a man who had been shot dead at the door of the tent in which he kept a small store: "the ball, of large calibre, entered the head behind and a little above the left ear, coming out of the forehead, tearing away the skull and spatting brains and blood on near objects"[2]. Addie Trevarthan married W. M. Williams at the Methodist Church in October 1889 just four days before Judge Kirby pronounced sentence on ten Apaches accused of murdering a cavalry officer, and just one week before sheriff Glenn Reynolds and his deputy were murdered by them on the way to the Yuma jail. And exactly a week after Cornish children had finished their Christmas parties, the one Apache found guilty of the actual shooting was executed in the Globe jail-yard by severing with one blow from an axe a rope tied to a great boulder of granite that dropped to the ground and shot his body to the top of the gallows.

Venturing out of Globe was a hazard, as the stage drivers knew only too well. George Kingdon from St. Austell, who sank the "K" shaft and was the superintendent of the Old Dominion, married the daughter of Charles Kinyon, in turn Pony Express rider and driver on the Butterfield Line: a useful combination of talent, it seems, for the wife safely drove the Cousin Jack out to Cananea in Mexico, where he managed a group of mines and amassed so considerable a fortune that he was known as the "millionaire boss". Venturing out in Globe could be equally dangerous for the unwary; it boasted fifty saloons that never closed; 150 prostitutes worked their way along the creek beds of North Broad; and as late as 1916 a girl, operating a soft-drink stand there, witnessed three murders in a single week[3].

As yet the Old Dominion showed no signs of advancing age. Under the control of its two Cornish superintendents, Samuel Parnell and Frank Juleff, who later left to become foreman for the Calumet-Arizona at Bisbee, by 1891 it had paid off all its accumulated debts. On 11 May 1889 a new smelter had been blown in and later a new hoisting

1 Ibid., 20 May 1882; 16 June 1888.
2 Ibid., 8 May 1889.
3 Miller, p. 136.

gear was installed to reduce costs still further; soon it was producing 7 tons of copper a day with one small furnace and seventy-five men, compared with the Copper Queen at Bisbee which was yielding 15 tons a day with three furnaces and 500 men[1]. By 1890 Globe was running a good second to Lake Superior because its ores were relatively cheaper to produce, in spite of the high cost of the coke which, at the Willcox railroad, was 16.75 dollars a ton and at Globe 46.75 dollars; but the ore was so pure that it cost only a little over five cents per lb to refine and could be sold in New York for just over seven cents. Producing one third of all the copper in Arizona, the Old Dominion finally went one stage further. Another smelter was blown in on 16 January 1892, so that its three furnaces could deal with a daily capacity of 185 tons of charge, ore and flux; and in the five years between 1888 and 1893 its total output of copper soared to 40,000,000 lbs[2].

The Old Dominion saved Globe from total collapse, for 1893 was the year of financial panic, leading to the repeal of the Sherman Silver Purchase Act and the end of silver mining in Arizona. Gone now were the days of the small-scale operators, for copper demanded the resources of big business with professionals in charge; and those who did not possess the skill and the experience of the Cornish now found themselves classed as "ten-day miners", willing to undertake any task for the price of a meal and a bed, ten days being the average time for the survivors. For them all copper came in the nick of time. In 1898 one shaft at the 1,200 ft level came across a remarkable ore body, sixty-five feet thick and hundreds of feet long. By 1900 it was reckoned that the streets of Globe were as busy at midnight as they were at noon, for this ninth greatest mine in the world, as it then was, was employing over 2,000 men, with 700 on each shift, until the new town of Miami had to be created in 1909 to cope with the phenomenal output of the new Miami Copper and Inspiration mines. And so at the Old Dominion the miners continued to follow the ore ever deeper, the "A" shaft, completed in 1904, finally touching bottom at nearly 3,000 ft. Then on 5 May 1930, after the price of copper had slumped to the unprecedented depths of five cents per lb, the mine closed for ever and its workers moved south to Bisbee and elsewhere.

The "Hoo-Dee", as the Old Dominion was known in almost every household in Cornwall, had its share of accidents, though the rate was generally low. Ernest Hosking, on a June day in 1913, was killed when

---

[1] *The Arizona Silver Belt*, 20 July 1889.
[2] Ibid., 9 August 1890.

he fell 85 feet down a shaft, and on Christmas Day James Pryor was brought out dead after a cave-in. The worst accident occurred in 1890 when John Isaac from St. Austell was timbering in the Interloper shaft alongside another Cornishman, Joseph Kinsman. The black mouth of the shaft was about 4 ft wide. Isaac, wanting to fetch a tool from the pump station, sprang across, as miners often did, but on this occasion he relaxed his caution of making sure whether the cage was working. He jumped at the very moment when the "shoe" of the upper cross-bar of the ascending cage met him and it carried him about 12 feet through two sets of timbers. His crushed body, when released, rolled over the side of the cage and fell 70 ft to the bottom of the shaft to suffer more indignities and disfigurements. A "worthy citizen of quiet disposition and gentlemanly demeanor", he was given one of the grandest funerals ever seen in Globe, more than 200 miners and their families following the hearse from the Methodist Episcopal Church to the Masonic cemetery[1].

Of his contemporaries little of substance seems to be known outside the files of tax returns and the register of voters. Philip Oates was the superintendent of the San Carlos copper mine, paid tax on real estate worth 200 dollars a year, went to Bisbee on the invitation of Douglas to inspect some coalfields at Sonora that the Copper Queen wanted to buy, and finally disappeared to Yucca "on mining business", taking with him Richard Edgcomb[2]. F. P. Drew herded 150 head of cattle worth 10 dollars each; and William Drew's possessions were his "residence, three quarters of a mile south of the court house, with five acres of land, four under fence, chicken house and granary, fire pump, variety of fruit, corral adjoining the grounds and good range for cattle"[3]. John Langdon, born in Hancock, after a time at Bisbee, Jerome and the United Verde Extension, became a master mechanic at the Old Dominion. From the north too came Frederick Hoar, trained at the College of Mines at Houghton. Manager of the Old Dominion in 1901, he practised as a consultant in that wildest of frontier towns, El Paso, and returned to Globe as general manager for the South-West Miami Development Company[4]. Straight from St. Just and then Cripple Creek, William Thomas Penrose for many years moved with authority as a shift boss at the Old Dominion until an accident finished him.

Some took the advice of an older generation and stayed in Globe

1 Ibid., 8 March 1890.
2 Ibid., 29 June, 20 July, 3 August 1889: 29 March 1890.
3 Ibid., 10 August 1889.
4 *Who's Who in Arizona* (1913).

only for two years because after that "the blood dried up", according to William Dyer of St. Austell. A clay-worker earning only 10 shillings a week, he decided on Globe simply because he knew the Bennetts there. With a through ticket for Globe which cost £14 he embarked at Southampton on the *Missouri* for New York. Blake's Hotel on Carson Street was the rendezvous for all Cornish in transit, and there he was given the best advice about the route by rail to New Orleans and then across Texas and New Mexico. Miss Carlyon's boarding house at Globe, he says, resounded with Camborne voices warning him against working for at least two months on account of the excessive heat; but this hardly seemed necessary, since there were at least 500 unemployed for reasons wholly unconnected with climate. These were the "ten-day" miners, and Dyer would have joined their ranks but for the fact that the superintendent of the Old Dominion was George Kingdon from Trethurgy, the pumpman came from Truro, and a Hoare from Bugle found him a menial job sweeping the tracks at the 600 ft level; and he was lucky for his wages were sixteen dollars a week. Nevertheless after two years he heeded the warning, returned to Cornwall for a wife, mined in Michigan and South Dakota and headed back for Cornwall before it was too late. But the majority of his contemporaries braved the heat and the depressions and died in Globe: Joseph Gundry from St. Blazey and Bingham Canyon; George Trenear who died of silicosis contracted in Johannesburg; John Kendall from Roche; John Trenarry; James Parkyn from Camborne; Elijah Quick who came by way of Idaho; a Benbow from Redruth; Richard Quick; Thomas Wakefield from Penzance; James Oatis Batten from St. Just; John Hicks from Callington; George Williams and Thomas Roberts from Redruth; James Henry Richards from Camborne; and Arthur Symons and John Williams from Truro[1].

Some, like Philip Hocking from Hayle, missed Globe and ploughed on to the dust of Morenci, just another Cornish town in a canyon where, as late as 1906, the only roads were trails trodden by pack mules; yet here a Mrs. Cocking came to join her husband from St. Ives and stood despondent in the main plaza as she regarded the desolation around her; but she adjusted herself to her unattractive surroundings and raised a family. Jerome was another of these remote Cousin Jack camps where Robert Mitchell and his four brothers supervised the smelter until he crossed the border at Sonora into Mexico, leaving behind him a town which, even until 1929, boasted a population of

[1] From the *Records of Naturalisation*, Globe.

15,000. Today it is but a ghost of its former self, its only link with the past a thriving historical society which emblazons on its official note-paper a picture of the miner's candlestick that the Cousin Jacks brought from Cornwall. But a very different fate awaited Miami.

Eight miles from Globe, it was deliberately created as a mining town. It represents the end of the long era of shift bosses, timbermen and engineers, and the contract system of wages that was peculiarly and wholly Cornish, for by then the way to exploit the low grade copper ores by handling them in immense quantities had been discovered. How immense is abundantly clear from the new mountains piled high upon each other from the open pit workings, no less than 90,000,000 tons of earth and 158,000,000 tons of tailings; this means that in fifty years 160,000,000 tons of ore have been blasted, grabbed and hauled to give a fantastic yield of 369,000,000,000 lbs of copper. The pioneers in this new method of extraction were the Miami Copper Company and the Inspiration Copper Company; together they fathered Miami Town, soon to be known as a "Pittsburg in copper", very different from its parent Globe which grew out of ranching and farming as much as out of gold, silver and copper. The property of the Miami Townsite Company was placed on the market on 11 October 1909; and by 1914 its population was 5,000 of whom 1,500 were employed with the two mining companies, their average wage ranging from 3.75 to 6.00 dollars a day for an eight-hour shift.

The Globe and Miami City Directory of 1913-14 lists a hundred Cornish householders, living at either Globe or Miami. The Gundrys, the Harrises, the Johnses, the Opies, the Richardses, the Thomases and the Williamses were especially thick on the ground; but among them is the bearer of a name that had been there from the begining, Thomas Pascoe; he was Vice-President of the Gila Bank and Trust Company. It would be invidious to draw distinctions in so distinguished a list; and there appear to be several families missing. For instance, the Renowden family of St. Ives established itself in Globe between 1909 and 1914; and John Angove from Camborne and Vicksburg, Mississippi, was working there 1912-1922. William Andrews from St. Agnes, a restless miner if ever there was one, bought passages for his family on the ill-fated *Titanic*, could not wait the two weeks until she sailed, and embarked on another boat for Globe, his family on this occasion grateful that "he had gotten in a hurry". Then there was the Hocking family and the tragedy that overwhelmed them one winter's night in 1916: "Lying weltering in his life's blood on the floor next to the bed on which reposed the body of his wife whom he had murdered, W. J. Hocking, a

Cornish miner, was found dead at his house in Miami at 1.0 a.m. Saturday"[1].

But there is one Cornish family that has proved to be unusually rich in its recollections of Miami, the Ellerys: and their saga has been told by William J. Ellery, who has died since this chapter was written. It was related in conversations with him in Penzance after a visit to the cottage in Camborne where he was born; in Grass Valley; in motels en route for Death Valley and at Lone Pine; and in his study at Monrovia in California among photographs of himself standing by the statue of Sir Humphry Davy at Penzance and of his friend Chief Talkalai, the Apache captor of Geronimo. This tall, spare and gaunt Cornish Arizonian and Californian would have been one of the first to admit that his life history is but typical of the thousands who have left no record, and that its telling is justified only as a reminder to posterity of their hardships in the struggle to survive and to provide a fuller life for their children.

His father, John Ellery, and his mother, Fanny Williams, had both known poverty as children in the dank and unwholesome streets of Redruth. Youths like himself were plentiful and cheap, and at the age of twelve he was working at Dolcoath mine as a "tool boy", carrying the long and heavy drills up and down the wet and slimy ladders in the shafts. The hours were long, the risks of a broken leg ever present, and the pay small. This drab existence in darkness and steamy heat was the only life he knew until he became a tin-dresser. At the age of twenty-six he married a milliner, but their hopes seemed blighted almost from the start. His wife's mother was an invalid with none of her family near to support her; one son had been sent to Brazil as an assayer; another had already left for America and was to prove an unexpected blessing later in Arizona; a third was helplessly deaf and dumb. W. J. Ellery was born in 1887; and John Ellery lost his position as a tin-dresser as the mines, one after another, closed down. Heavily committed to his family responsibilities, John Ellery therefore made the decision to emigrate, leaving behind his wife to care for the two invalids and their two months old son.

For ten lean years, in Michigan and Colorado, he lived and worked, regularly sending money home and improving his position, but never pressing his wife to solve their problem by putting her mother away in some Poor Law institution, as she might well have done. But in 1896 she died and so Mrs. Ellery and their son, who had no memory of his

---

[1] *The Arizona Silver Belt*, March 1916.

father, were at last free to make the long journey to Guston, a silver camp, long since disappeared, on a mountain top in the Rockies between Ouray and Silverton, where the Denver and Rio Grande railroad clung to granite crags and spanned canyons and gorges over matchstick bridges. But their mountain idyll and their second honeymoon did not last long, for in 1897 the price of silver collapsed and the mine closed. So once more they were on the move, this time on a southbound train to Globe, since John Ellery knew a Redruth miner there, Milt Williams, who promised him work, for the Old Dominion was short of hands. W. J. Ellery remembers that it was not an easy journey, for his mother was nursing his three months' old brother: from Guston to Silverton by buckboard; by rail to Geronimo; finally a ride of sixty miles across the desert on a bumping stage through the Apache reservation with an overnight stay in the Indian town of San Carlos. But after six years the "old debil" again tapped them on the shoulder, there was a recession in 1903 and John Ellery found himself unemployed.

However, his father's brother-in-law, Edwin Nettell, was a mine captain in Grass Valley, and that connection was a sufficient guarantee of work if it was available. Having to leave Globe could not have come at a worse time, for his wife was ill and there were household debts, not unusual when miners were paid monthly. But the owner of the Old Dominion store, George W. P. Hunt, later Governor of Arizona, told them to pay when they could. Work at Grass Valley lasted only two years for John Ellery's mining life was suddenly brought to an end by a stroke. So now it became the turn of the son to make the sacrifice to support his family when he had completed only two years of his high school education. Delivery boy and assistant in Grass Valley shops, earning 50 dollars a month for a twelve-hour day, he nevertheless continued his education at night at the local business college, studying accountancy. Then came a slice of luck when his mother's brother in Globe offered him a post as book-keeper at 100 dollars a month. This new rate seemed princely until he remembered that he now had to support two households: his own lodgings in Globe cost 30 dollars and he remitted 60 dollars a month to his parents in Grass Valley. Then he was invited to be a book-keeper for the National Bank of Globe and, though it meant a reduction in salary, accepting proved to be the right decision.

In 1909 the bank decided to open a branch on Miami Flat in anticipation of the new mining town, where it was expected that there would be more saloons than shops and more miners and steelworkers than churchmen; and a steady hand was needed to pay out to the rough and ready workers. S. W. J. Ellery was chosen as the first cashier. Since there was no building that could be rented, he bought a frame store, 10 ft square, for 40 dollars, made a grill out of chicken wire, daubed the name of the bank on it with paint and waited for business. On 8 February 1909, the day before his chicken-coop bank opened, he might have been seen leaving Globe in a wagon driven by the "black pioneer", Alvin Booth, the descendent of a negro slave from Texas, with bags at their feet containing 50,000 dollars in cash, protected by one pearl-handled revolver between them. The first customer, who wanted the distinction of being "No. 1" on the books, was the owner of the bar in the red light district; he opened his account with the not inconsiderable receipts of the previous night. There was no strong room; the books had to be deposited each night in the safe over at FitzPatrick's saloon; and Ellery had to be both janitor and cashier, sleeping on an army cot with a .45 revolver under his pillow. Though he never had to use his weapon, for a young man of twenty-three these were anxious first days, for there was as yet no jail in Miami, drunks were locked in railroad box cars, and the magistrates imposed fines as they sat astride a barrel on the sidewalk outside the Palace Saloon.

From now onwards the Ellerys never looked back. As Miami was growing quickly and houses were difficult to find, with their carefully accumulated capital they built one for themselves and several others for rental until they had acquired a small but steady income. Then Ellery married, wisely choosing a wife who came of sound pioneer stock from Wisconsin; and their children represented the new America that was born and nurtured on the harsh frontier experiences of the old. One son graduated at the California Institute of Technology and founded his own firm of consulting engineers; another qualified at the University of California as a Doctor of Optometry; and a daughter read Social Science at Stanford. Later their father was promoted President of the Miami Bank and his brother Vice-President. In 1922 W. J. Ellery moved to California to take over the position of Vice-President of the Security First National Bank of Los Angeles and his brother in Miami distinguished himself as a financial expert, for during the depression, Governor Hunt invited him to be State Superintendent and Inspector of Banks. That the son of a Cornish miner should undertake such a responsibility at the time of the financial chaos in 1929 is probably unique; and what is quite astonishing is that there was no banking experience in the family at all: nothing but a rugged road of meagre resources, hardships and disappointments that ran all its twisted way from Redruth and Camborne.

The Cornish banker knew everyone in Miami for they traded their cheques over his counter. But none was more welcome in Ellery's office than the Apache Chief Talkalai who, wearing his feathered headgear or an old battered sombrero, would chat the hours away, smoking a big black Owl cigar specially reserved for him. Not without reason he would conduct himself with as much dignity as if he were the President himself; raised on the San Carlos Reservation, he had been Chief of Scouts for the army under three generals: Crook, Miles and Howard; and for twenty-one years Chief of Police on his reservation. Few of the Cornish miners seeing him shuffle out of Ellery's bank realised that he had saved the life of Colonel John Clum when, with a detachment of Apache police, they had gone to Ojo in New Mexico, near the headwaters of the Gila River, 400 miles by trail from San Carlos, to capture Geronimo one April day in 1887. Later generations forgot how he had once been the guest of President Cleveland in the White House; and that his finest hour was in March 1930 when he was the guest of honour at the opening of the Coolidge Dam. Standing alongside the President, Colonel Clum and Will Rogers, he heard the guns boom and saw the waters rolling over the acres where his ancestors had once hunted; but the occasion proved too much for the old man, and they buried him in the white man's cemetery in Globe on Pascoe Hill, not far from the Cousin Jacks who also had helped to tame the tiger of Arizona.

By then, for Cousin Jacks almost everywhere, their sun had almost completed its descent. Gone were the days when they could step off the boat at New York with little money and no passport, knowing that, if they could ride rough as far as Globe, "Ma and Pa" Shugg would take them into their boarding house; and when W. J. Ellery went to school with the son of Hampton Blevins, one of the "hash riders" who was shot down by the Tewksbury "boys" in that bloody feud over the rights of sheepherders with the Grahams in the Tonto Basin that almost annihilated the two families. The old frontier his father knew was now halted before the huge gaping mouth of the open-cast pit, where the mechanical slaves of the modern industrial world grabbed and clawed at the low-grade ores, and management and men temporarily lost sight of each other. Strikes for union recognition embittered both; those of 1903 in the Clifton-Morenci district, when the National Guard and the Arizona Rangers were called out to restore order and protect property, were serious enough; but those of 1915-17, caused by the appearance of agitators from the militant Industrial Workers of the World, confused real economic grievances with left-wing political activity. In the summer of 1917 more than 1000 miners from Bisbee were hustled at bayonet point into a special train and driven into the desert of New Mexico, where, without adequate supplies of food, water or shelter from the blistering heat, they were abandoned by their guards and left to fend for themselves. In Globe the miners refused to fly the Union flag over their Union Hall.

Though the surviving Cornish were reluctant to join the struggle and were more disposed to drive the I.W.W. agitators out of town, they knew that the day of the hard-rock miner was over, remembered only by the old-timers as they sat on the high steps of the Court House beneath the plaque to their friend Governor Hunt, whose personal influence and intervention alone had prevented widespread violence and bloodshed. Labour relations did improve in the 1920's but could not stop the depressions of the magnitude of 1929, when both managers and men found themselves without work. Oddly enough, it was then that pamphlets were published throughout Arizona advising prospectors how to search for gold and silver; and the skills that had once been locked in Cornish fingers and fists, the arts of drilling, blasting and timbering, were then no more than tables of instructions for destitute and inexperienced amateurs. *Sic transit gloria Cornubiae.*

Accidents, 63, 67, 68, 69, 93, 96, 97, 102, 107, 111, 120, 123, 124, 130, 132, 140, 142, 143, 151, 155, 162, 164-5, 168, 173, 178, 181, 183-207 *passim*, 217, 219, 225, 241, 246, 247, 261-2
Alaska, 78, 107, 109, 182, 243
Albuquerque, 120, 248
American Civil War, 19, 24, 31, 35, 37, 40, 60, 80, 105, 120, 122, 125, 139, 146, 216, 221, 229, 236, 246, 248, 252
Apex Law, The, 243-4
Arbuckle, Clyde, 11, 80, 81
Arizona, 10, 15, 75, 76, 109, 130, 131, 177, 180, 181, 182, 233, 237, 247, 248-269
Assaying, 52, 56, 77, 151, 182, 265
Astor, Jacob, 29
Australia, 60, 75, 109, 155, 182, 204, 228, 237

Bad Axe, Battle of, 29
Baker, John, 32-35, 45-46
Baker, John Earl, 32, 34
Bands, Cornish, 75, 77, 101, 107, 132, 141, 152, 178
Barkerville, 15
Bear Flag Republic, 50, 83
Black Hawk War, 29
Boarding Houses, 55, 72, 77, 82, 104, 117, 119, 138, 156, 163, 173, 174, 178, 179, 188, 228, 231, 245, 246, 253, 256, 263, 268
Bolivia, 19, 86
Bonanza, The Big, 193
Borthwick, J. D., 59
Boston, 119, 126, 127
Boxing, 75, 101, 104, 194-5, 243
Brazil, 225, 265
British Columbia, 15, 85, 86, 107, 140, 176, 208, 218, 221
Buffalo, 22, 44
Bullock, Richard (See Deadwood Dick) 9, 27, 149
Bulmore, L. E., 11, 80, 81, 82

California, 13, 30, 50-113, 125, 130, 146, 150, 155, 182, 185, 189, 197, 203, 208, 217, 221, 233, 248, 250, 254, 256, 259, 265
Canada, 45, 118, 119, 232, 245
Cardiff, 259
Carson, Kit, 50, 155
Cartee, General Lafayette, 228
Cheyenne, 161
Chicago, 72, 114, 132, 143, 161, 259
Chile, 19, 56, 86, 119
China, 147
Chinese, 55, 70, 188, 191, 192, 231, 247
Choirs, Cornish, 58, 75, 77, 78, 94, 101, 104, 105, 106, 108, 122, 124, 125, 132, 154, 161, 178, 245

Citizenship, American, 68, 77, 93, 126, 148, 192, 258
Clarke, W. A., 238, 238, 241, 243, 244
Colorado, 74, 76, 77, 107, 125, 132, 151-182, 183, 208, 217, 219, 231, 250, 256, 265, 266
Comments on Cornish, 67, 68-9, 70, 72, 77, 85, 90, 94, 106, 109, 116-7, 120-6, 127, 129, 132, 137, 139, 142-3, 151-2, 154, 156, 157, 160, 161, 162, 164, 166, 182, 186, 189, 203, 221, 225, 227, 230, 247, 253
Companies, Mining and Smelting, 56, 59, 62, 63, 64, 69, 74, 77, 80, 83, 84, 94, 101, 102, 111, 115, 116, 117, 118, 119, 120, 126, 128, 129, 132, 137, 138, 148, 153, 159, 160, 161, 175, 178, 181, 186, 188, 189, 193, 217, 219, 225, 226, 237, 238, 247, 250, 254, 257, 264
Comstock, Henry P., 185
Comstock Lode, 63, 72, 106, 183, 188-207 passim, 226
Congressmen, Cornish, 75, 131, 132, 137, 175, 216, 230, 242
Connecticut, 85, 229, 247
Cooper, Fennimore, 98, 209
Copeland, Louis Albert, 28, 29, 30, 46, 47, 59
Council Bluffs, 61
Cuba, 77, 93, 11, 105, 119
Culture, of Cornish emigrants, 25, 26, 30, 31, 35, 36, 40, 41, 48-9, 58, 70, 73, 75, 77, 94, 101, 102, 123, 142-3, 151-2, 156, 178, 194

Dakotas, The, 104, 130, 131, 148-50, 182, 203, 222, 230, 259, 263
Daly, Marcus, 238, 241, 242, 243
Day, Sherman, 84
Deadwood Dick (see Richard Bullock) 9, 27, 149
Death rates, 68, 96, 98, 160, 176-7, 184-5, 253
Disease, 44, 45, 93, 108, 118, 122, 126, 155, 160, 162, 174, 176-7, 181, 198, 233
Divorces, 72, 98, 198
Dodge, Colonel Henry, 29
Drilling Contests, 75, 81, 101
Drinking, 67, 73, 74, 85, 101, 105, 119, 121, 123, 128, 141, 162, 184, 195, 196-7, 199, 227, 236, 250
Dyer, Dr. Frank, 143-4

Earp, Wyatt, 251
Education, Cornish achievements in, 36, 40, 41, 49, 55, 64, 70, 72, 73, 75, 76, 77, 82, 86, 87, 91, 93, 97, 98, 101, 102, 105, 106, 107, 109, 115, 116, 122, 123, 128, 130, 131, 132, 137, 140, 142, 144, 145, 146, 149, 150,

152, 156, 175, 191, 194, 210-15, 216, 217, 227, 230, 234, 237, 256, 264, 266, 267

Elections, 28, 80, 88, 101, 122, 123, 162, 221, 226, 243

Elgin (Iowa), 36, 38

Ellery, W. J., 265-9

Emigration, Reasons for, 20, 39, 67, 116, 126, 130, 143, 154, 230-1, 233-4, 237

Emigration, Statistics of, 9, 13, 14, 19, 20, 21, 29, 50, 67

Entertainments, 94, 96, 101, 105, 108, 123, 141, 191, 226, 228, 260

Episcopalian Church, 40, 55, 104, 122, 143, 157, 184, 190, 194, 207, 243

Erie Canal, 22, 44

Explosives, 69, 72, 108, 119, 124, 151, 164, 165, 195, 202, 231, 237, 242, 243

Families, Cornish emigrant,
Adams, 68 Allen, 176 Andrewartha, 225 Andrews, 250, 264 Angove, 75, 143, 198, 232, 264 Argall, 58, 76, 93, 258 Arthur, 140, 216 Axford, 151 Bailey, 168 Barkel, 124 Barling, 176 Barnes, 124, 229 Barnett, 201 Barrett, 93, 179, 198 Barron, 107 Bartle, 93 Basset, 186 Bastian, 138 Batten, 263 Bawden, 69, 107, 138 Benbow, 263 Bennallack, 64, 68, 74, 76 Bennet, 263 Bennett, 58, 63, 120, 201 Bennetts, 124, 137, 245 Berry, 186 Berryman, 46, 74, 86, 93 Betallack, 68 Bettis, 63 Bickford, 156 Billings, 202 Bishop, 92, 102, 168, 192, 246-7 Blackwell, 154-5 Blewett, 225, 256 Blight, 186, 193 Bluett, 62, 107, 202 Bone, 32-35, 77 Bosustow, 42 Bottrell, 144 Bowden, 92, 168, 176 Bray, 58, 144, 176, 242 Brent, 156 Buckley, 186 Bullock, 27, 63, 149 Bunney, 92, 144 Burden, 150 Burgan, 137 Burge, 138 Caddy, 55, 57, 232 Cann, 148 Carkeek, 107 Carlis, 168 Carlyon, 91, 144, 263 Carne, 104, 107 Carter, 74 Carthew, 58 Chalmers, 42 Champion, 200, 203 Chapman, 144 Chegwidden, 55 Chellow, 179 Chirgwin, 258 Clemo, 55, 144 Cliff, 137 Clyma, 28 Cock, 76 Cocking, 245, 263 Colenso, 144 Collick, 140 Collins, 115 Colliver, 64, 87, 102 Combellack, 144, 216 Connibear, 145 Cook, 201, 216 Coombes, 73 Copeland, 28, 29, 30, 46, 47, 59, 191 Cord, 63 Cornish, 92, 93 Coulson, 52-3 Cowling, 250 Cox, 32, 144, 201 Crapp, 32 Crase, 64, 67, 75 Craze, 55, 58, 69, 186, 198 Cudlip, 142 Curnow, 55, 58, 92, 106, 107, 176, 182, 191, 258 Dale, 40 Daley, 198 Daniell, 137 Darke, 126 Davey,

137, 225, 226, 245 Davy, 227 Dennis, 138 Dingle, 144, 182 Doidge, 92, 94 Drew, 67, 102, 223, 229, 262 Dunstan, 92, 107, 137, 176, 216, 244 Dunstone, 168 Dyer, 143-4, 263 Eddy, 63, 92, 94, 178, 186, 200, 201, 258 Ede, 107 Edgcomb, 262 Edwards, 58, 92, 115, 123, 126-7, 144, 229 Ellery, 265-9 Ennis, 72 Eva, 144 Evans, 200 Faull, 230, 258, 259 Ferrel, 250 Fitzsimmons, 25, 104 Floyd, 173 Foxwell, 35-39, 40, 42-45, 209 Francis, 216 Furze, 148, 232 Gale, 156 Gartrell, 39 Geach, 87, 101, 144 George, 40, 42, 46, 58, 62, 69, 73, 75, 93, 124, 174, 177 Gilbert, 92 Glasson, 74, 76, 186 Glendenning, 46 Gluyas, 203 Godfrey, 115 Goldsworthy, 40, 64, 75, 140, 192, 201 Green, 116 Gregor, 30 Grenfel, 32, 115, 216 Grey, 91 Gribble, 69, 107, 192, 254 Grigg, 14 Gundry, 48, 107, 263, 264 Hall, 124, 130-1, 149, 178 Hancock, 41, 42, 46, 93, 186, 258 Harper, 86 Harris, 35, 40, 55, 92, 93, 132, 137, 148, 168, 229, 264 Harry, 81, 87, 91-2, 94, 102-3 Hart, |250 Harvey, 141 Hawe, 45 Hawken, 107 Haws, 229 Hay, 42, 43, 60 Heather, 55 Hender, 107 Hendra, 140 Henwood, 92, 137, 176, 181 Hendy, 32 Hichens, 144, 145 Hicks, 177, 263 Hitchens, 144 Hoar, 118, 132, 262 Hoare, 263 Hocking, 75, 144, 176, 192, 245, 257, 264 Hodge, 93, 98, 107, 109, 127 Hodges, 196 Hollow, 58 Holman, 98 Hooper, 32 Hoskin, 64, 168 Hosking, 141, 176, 261, 263 Hoskins, 45, 69, 107, 146, 162, 174, 178, 226 Hugh, 250 Humphrey, 38, 44 Hunkin, 184 Inch, 106, 107 Ingram, 55 Isaac, 262 Ivey, 42, 48, 144 Jacka, 92, 137, 183 James, 29, 37, 38, 39, 40, 44-5, 58, 63, 75, 123, 140, 144, 209-215, 247, 258 Jenkin, 74 Jenkins, 62, 63, 77, 183, 186, 201, 232 Jennings, 88, 94, 102, 103, 124, 225, 258 Jewell, 72, 75, 203, 228, 256, 258 Johns, 92, 107, 115, 180, 188, 202, 264 Jollif, 32 Jose, 27, 104-6, 204 Juleff, 260 Keast, 144 Kendall, 149, 263 Kent, 230 Kessel, 93 King, 26, 141, 151, 259 Kingdon, 260, 263 Kinsman, 69, 253, 259, 262 Kinyon, 260 Kissell, 237 Kitt, 176 Kitto, 64, 225, 226 Kittoe, 46 Knapp, 229 Kneebone, 141 Knight, 252, 255 Laity, 202 Langdon, 262 Lanyon, 101, 167 Lawry, 156 Lean, 32, 137 Levering, 162 Lewarne, 96 Liddicoat, 72, 198 Lidicote, 200 Lobb, 107, 255, 259 Lory, 41, 42, 210 Lugg, 14, 40, 42

Luke, 105, 186 Mantle, 245 Martin, 40, 92, 144 Masters, 118 Matthews, 92, 96, 215, 216 Maynard, 28 Millet, 257 Mitchell, 56, 62, 63, 72, 73, 74, 76, 107, 154, 162, 179, 186, 192, 198, 201, 204, 223, 227, 229, 258, 263 Morcam, 58, 149 Morrish, 200 Moyle, 15, 36-7, 39-44, 46, 107, 174, 176, 201, 210, 233-5, 257 Murton, 246 Nance, 258 Neal, 47 Nettel, 55-6, 266 Nettle, 189 Nicholas, 144 Nichols, 67, 106, 107, 132, 143, 148, 174, 217, 258 Ninnis, 199, 204-5 Noall, 200 Northey, 189 Nye, 62 Oates, 46, 63, 93, 179, 192, 244, 258, 259, 262 Oatey, 144 Odgers, 46, 69, 92, 179, 198, 201 Olds, 223 Opie, 56, 192, 251, 264 Osbiston, 162 Osborne, 122, 123 Owsley, 245 Painter, 48 Parkin, 251 Parkyn, 263 Parnell, 260 Parsons, 48 Pascoe, 55, 64, 68, 69, 75, 76, 92, 109, 110, 112, 141, 144, 188, 192, 219, 229, 251, 256-7, 258, 259, 260, 264 Paul, 123, 176, 258 Paull, 118 Paynter, 75, 245 Pearce, 45, 52, 59, 85, 86, 91, 92, 93, 146-8, 153-4, 164, 176, 201, 245 Pellow, 198 Pellowe, 144 Penberthy, 122, 123, 177, 192 Peneluna, 176 Pengelley, 55, 144 Pengilley, 45, 140 Penhallow, 14 Penpraze, 165 Penrose, 77, 181, 198, 199, 262 Perrin, 62 Peters, 58, 63 69, 73, 75, 107, 130, 193 Phillips, 29, 48, 52, 144 Pierce, 101, 179, 201, 223, 253 Polglase, 62, 63, 132, 137, 204, 230 Polkinghorne, 58, 144, 188, 192, 201 Pomeroy, 179 Pope, 186, 192, 203, 245 Prideaux, 31, 216 Prisk, 58, 76, 168, 176, 225 Prouse, 144 Prout, 93, 144, 176, 229 Prowse, 237 Pryor, 92, 144, 262 Quick, 74, 263 Ralph, 58, 58, 92 Rapson, 124, 185 Rawlins, 119, 137 Reese, 68 Renowden, 264 Reseigh, 92 Retallack, 185 Retallic, 132 Reynolds, 72, 148, 236 Richards, 29, 30, 46, 68, 69, 107, 124, 132, 137, 138, 139, 140, 144, 149, 156, 167, 177, 186, 192, 263, 264 Rickard, 59, 124, 144, 167, 182 Roach, 196 Robert, 73 Roberts, 263 Robins, 92, Rodda, 18, 64, 68, 75, 216-7, 229 Roe, 186, 216, 252 Rogers, 73, 144, 185 Rolfe, 165 Rosewarren, 107 Rosevere, 193 Roskilley, 42 Rowe, 46, 55, 67, 72, 75, 87, 92, 109, 139, 141, 144, 181-2, 186, 188, 195, 223, 225, 226, 227-8 Rowse, 62, 186 Rule, 30, 46, 63, 73, 93, 161, 186, 196-7 Rumphery, 48 Rundle, 141 Scaddon, 223 Scoble, 51 Scovell, 226

Searle, 26 Semens, 124 Semmen, 167, 173 Shephard, 40, 41, 42, 43, 46, 210 Shugg, 144, 268 Simmons, 141 Sims, 232 Sincock, 46 Sinnott, 199 Skewers, 196-8 Skewes, 40, 42, 43, 46 Skewis, 45, 60, 61, 62 Sleeman, 48, 115, 141 Smith, 115 Snell, 144, 179 Sobey, 52 Spargo, 180, 186 Stacey, 226 Stanway, 139 Stenlake, 51 Stephens, 46, 144, 186, 192 Stevens, 63, 69 Stocks, 76 Stoddart, 124 Stratton, 132 Strike, 144-6 Strongman, 61 Symons, 263 Tallon, 251 Tamblyn, 41, 58, 144 Tangye, 46 Teague, 198 Terrell, 119 Terrill, 217-8, 232 Tierney, 74 Tink, 50 Tippett, 141 Thomas, 39, 42, 43, 44, 55, 56, 74, 75, 76, 77-9, 116, 178, 188, 190, 219-221, 225, 235, 237-8, 244, 264 Tonkin, 87, 92, 98, 102, 138, 141, 225 Toy, 176, 229 Trathan, 161 Trebilcock, 32, 46, 137, 140 Tredinnick, 55 Tredrea, 48, 75 Tregaskis, 111-2, 127, 223 Tregloan, 46, 188, 198 Tregonning, 30, 46, 74, 92, 93, 94, 101, 118 Trelawn, 118 Treloar, 31, 48, 62, 115, 241 Tremaine, 118, 193 Trembath, 74, 137, 188, 196, 201 Trenberth, 64 Trenear, 263 Trengrove, 92, 107, 258, 259 Trennary, 259, 263 Trerew, 258 Tresidder, 112, 113, 123 Trestrail, 46 Trethewey, 118, 123 Trevithick, 9 Truan, 161 Trudgeon, 51 Truscott, 190 Trevarthan, 92, 125, 258, 260 Trevathan, 15, 255 Trevillian, 118, 138, 179, 185, 194-5, 251, 252, 258 Trevithick, 68 Trevorrow, 93 Trewarkus, 198 Trewartha, 46, 122, 125, 140, 256, 258 Treweek, 48 69, 140, 186, 190 Trewen, 257 Trewhella, 145, 202 Treworjy, 179 Trewyn, 32 Trezise, 74, 156, 167, 192 Trezona, 72 Tucker, 141 Turrell, 163, Tyacke, 229 Uglow, 32 Uren, 50 63, 137, 141 Varcoe, 87, 92, 94, 102 Venning, 51 Vial, 191 Vine, 168, 242, 251 Vivian, 29, 45, 123, 177 Vyvyan, 40 Wakefield, 259, 263 Warren, 115, 227, 229 Waters, 174 Webb, 245 Wedge 178 Werry, 75, 230-2 Whale, 140 White, 148 Whittle, 138 Wicks, 199 Wilcox, 115 Willey, 203 Williams, 75, 122, 144, 145, 157-8, 165, 167, 168, 173, 191, 195, 200, 230, 236, 245, 263, 264, 265, 266 Willoughby, 93 Wilton, 230 Winn, 32, 186 Woolcock, 113 Worth, 144 Wyatt, 144

Farming, Cornish, 30, 31, 32, 34, 35, 36, 37, 38, 40, 41, 45, 64, 70, 75, 86, 92, 94, 95, 116, 131, 175, 203,

210-15, 229, 230, 253, 256, 257, 262
Fever River, 28
Fisher, James, 116
Folk-lore, 101, 132, 152, 154, 241-2
Fort Hall, 212
Fraser River, 15, 221, 232
Frémont, John Charles, 50
Friendly Societies, 48, 87, 96, 97, 98, 101, 105, 156, 158, 184, 200, 222, 225, 226, 228, 229, 256
Gambling, 101, 123, 196-7, 250
Georgia, 152, 156, 225, 234
Geronimo, 109, 265, 268
Gilbert, John, 15
Gold Hill (North Carolina), 15
Gratiot, Colonel C. H., 30
Graveraet, Robert, 138
Grosch, Allen and Hosea, 185
Hamlin, Mr. & Mrs. H., 11, 104, 244
Harte, Bret, 54
Hearst, George, 241, 243
Heinze, Frederick, 243-4
High-grading, 56
Hill, N. P., 153, 160, 162, 163
Hobart, Henri A., 120-6 *passim.*
Hornblower, Jonathan, 9, 14
Horse-whim, 48, 107
Houses, Cornish, 47-8, 55, 74, 82, 94, 95, 116, 119, 123, 128, 130, 141, 153, 155, 156-7, 188, 194, 198, 235
Hudson's Bay Company, The, 209, 214, **228**
Hudson River, 22, 44, 195
Hunt, George W. P., 266, 267, 268
Hydraulic mining, 51, 75, 103, 222
Idaho, 9, 68, 72, 76, 111, 140, 189, 208, 219-232, 233, 245, 248, 250, 251, 263
Illinois, 14, 144, 248
Illinois River, 60
Independence Rock, 212
Indiana, 132, 248
Indians, 29, 30, 47, 55, 75, 83, 96, 103, 107, 108, 109, 110, 113, 115, 116, 131, 138, 156, 161, 174, 176, 186, 188, 210, 211-15, 225, 235, 248, 249, 251, 255, 256, 257, 260, 265
Insurance, 69, 95-6, 202
Investment, Cornish, 64, 68, 73, 74, 76, 77, 96, 125, 126, 127, 165, 178, 181, 186, 188, 229, 253, 257, 258, 260
Iowa, 36, 61, 210
Irish, 70, 101, 119, 138, 139, 162, 165, 195, 202, 241, 242, 243
Irving, Washington, 209
James, Samuel, 209-215
Johannesburg, 56, 98, 263
King, Alfred Castner, 151-2
Laramie, 26, 211
Lawmen, Cornish, 29, 55, 65, 67, 75,

76, 138, 139, 140, 157-8, 175, 229, 257
Lead mines (Wisconsin), 46, 47, 48
Learned Societies, 15, 19, 26, 51, 55, 56, 80, 81, 106, 153, 223, 236, 264
Lewis and Clark, 209
Literature, the Cornish in, 116-7, 142-3, 151-2, 241-2
Liverpool, 21, 45, 233
Loomer, Harlin G., 32, 35
Machinery, Mining, 48, 52, 54, 55, 59, 61, 62, 63, 67, 68, 71, 74, 80, 81, 82, 85, 90, 91, 94, 102, 103, 108, 109, 118, 119, 127, 128, 132, 137, 142, 153, 160, 161, 178, 186, 188, 189, 198, 201, 204, 225, 226, 231, 232, 250, 252, 257, 258-9
Malaya, 19
Manley, William Lewis, 110
Marshall, James, 51
Mayors, Cornish, 55, 75, 157, 175, 245
Maryland, 86, 231
Medical care, 74, 95-6, 98, 120, 128
Mexican Miners, 80-103, 250
Mexico, 15, 50, 51, 58, 59, 60, 84, 87, 93, 212, 217, 225, 252, 260, 263
Methodism, 24, 25, 30, 32, 35, 36, 40, 44, 64, 101, 115, 116-7, 122, 130, 142-3, 144, 145, 146, 149, 215, 225, 229, 230, 242
Methodist Episcopal Church, 29, 30, 35, 40, 41, 42, 43, 46, 48, 55, 58, 73, 80, 82, 92, 97, 98, 106, 112, 115, 141, 142, 150, 154, 156, 157, 178, 184, 190, 194, 201, 207, 257, 258, 260, 262
Michigan, 9, 10, 55, 56, 87, 93, 107, 109, 114-150, 151, 173, 174, 181, 183, 230, 246, 250, 251, 252, 258, 259, 263, 265
Military Service, 29, 31, 64, 115, 122, 177, 234
Mines:
CORNWALL: Balleswidden, 85 Bolingey, 237 Botallack, 219 Dolcoath, 18, 252, 265 Fowey Consols, 18 Gwennap Consols, 18 Kit Hill, 116 Prince of Wales, 148 South Crofty, 115 Tincroft, 56, 115 Wheal Busy, 126, 232 Wheal Charlotte, 77 Wheal Clifford, 196 Wheal Kitty, 115 Wheal Lushington, 77 Wheal Towan, 77 Wheal Vor, 91, 132
ARIZONA: Andy Campbell, 257 Carrie, 257 Copper Queen, 237 250, 261 Golden Eagle, 257 Golden Nugget, 257 Heintzelan, 254 Hoosier, 257 Independence, 257 Illinois, 257 Interloper, 259 King of Arizona, 253 Lost Gulch, 256 Mack Morris, 257, 258 Magma, 254 Moffat, 257 Old Dominion, 251, 252, 257, 258-9,

260-2, 261 Petaluma, 256 Pierce, 253, San Carlos, 262 Silver King, 254, 255 Silver Queen, 254 Stonewall Jackson, 257 Tacoma, 257 Tiger, 232 True Blue, 257 Vulture, 254

CALIFORNIA: Black Lead, 74 Black Oak, 107 Boston Flat, 64 Bullion, 74 Calico, 108-110, 130 Cedar Quartz, 71 Central North Star, 76 Champion, 58 Cornish, 68 Draper, 107 Empire Star, 54, 69, 70, 71, 72, 76 Erie, 74 Eureka, 64 Gold Hill, 61, 62 64, 67 Gold Point, 76 Harvard, 103 Howard Hill, 71, 74 Hues Hill, 64 Hueston Hill, 64 Idaho Maryland, 55, 58, 69, 71, 204 Imperial, 76 Independent, 107 Landers, 107 Lucky, 67 Massachusetts Hill, 61, 64, 74 New Almaden, 72, 79 103, 106, 107, 109, 128, 203, 258 New York Hill 63, 76 North Banner, 76 North Bloomfield, 88 North Star, 54, 64, 71, 75, 76 Oakland, 74 Ophir, 69 Peabody, 74 Pennsylvania, 54 Phoenix, 76 Sebastopol, 74 Sierra Buttes, 74, 78 Slate Ledge, 5 Sneath and Clay, 54 Soulsby, 104, 106, 107 South Star, 68 Town Talk, 64 Wheal Betsy, 64 Whitlock, 112 Yellow Jacket, 69

COLORADO: Alps, 168, 219 Bobtail Consolidated, 166 Briggs, 163 Calhoun, 168 California, 167, 182 Caribou, 176-7 Carr, 168 Cook, 168 Corydon, 158 Cripple Creek, 180 Enterprise, 182 Fisk, 164 Georgetown, 177-9 Gregory Bobtail, 163, 167, 168, 173, 174 Gregory Consolidated, 156, 158, 161 Hunter, 165 Illinois, 162 Kansas, 182 Kent County, 182 Leadville, 179-80 Leavitt, 163 North Star, 162 Ophir, 165 Original, 168 Parnlee, 163 Rising Sun, 162 Running Lode, 168 Russel Gulch, 156, 231 Silverton, 176 Sleepy Hollow, 168 Smith, 163 Telluride, 219 Wood, 153 Yankee Girl, 182

IDAHO: Afterthought, 229 Black Jack, 229 Buffalo, 225 Bunker Hill, 232 Chariot, 228 De Lamar, 228 Dewey, 228, Dollar, 229 Galena, 230 Ida-Elmore, 225, 226 Minnie Moore, 230, 231 Poor Man, 228 Rocky Bar, 223-8 Silver City, 228-9 Standard, 219 Sunshine, 232 Tiger, 219 War Eagle, 228 Wilson, 228

MICHIGAN: Ahmuk, 126, 127, 137 Albion, 126 Allouez, 137 Arcadian, 132 Atlantic, 132, 137 Boston, 118 Calumet, 127, 131, 132, 137, 138 Centennial, 137 Central, 116, 125, 130, 132, 137 Champion, 140-1, 145 Chapin, 141, 142, 150 Cleveland, 139 Cliff, 119, 120-6, 127, 130, 137, 229 Columbian, 126 Concord, 132 Copper Falls, 130, 137 Douglas, 132 East, 140 Edwards, 132 Franklin, 126, 137, 138 Hecla, 127, 131, 137 Humboldt, 123, 139 Huron, 132 Iron Mountain, 139, 141, 142 Keystone, 140 Lake Angeline, 139, 140 Minong, 137 Mohawk, 127, 132 New York, 139 North American, 137 Osceola, 126, 137 Pewabic, 136, 137, 138 Phoenix, 137 Quincey, 126, 132, 137, 149 Seneca, 127 St. Mary's, 132 Superior, 132 Tamarack, 126, 127 Wheal Kate, 126

MONTANA: Anaconda, 238, 243 Belmont, 232 Black Rock, 245 Cobra, 243 Colusa-Parrott, 245 Granite Mountain, 247 McDermit, 243 Minnie Healey, 243 Moulton, 245 Mountain View, 245 Neihart, 245 Never Sweat, 243 Ontario, 241 Original, 245 Pennsylvania, 244 Poser, 245 Rarus, 243 Speculator, 245 Steward, 247 Stuart, 245 Tramway, 245 Travona, 245

NEVADA: Belcher, 193, 194, 202 Central, 188 Chollar Potosi, 188, 191, 193, 200 Consolidated Virginia, 194, 201 Crown Point, 190, 194, 201 Emma Nevada, 204 Gould and Curry, 193, Hale and Norcross, 194, 200 Hope, 192 Imperial, 202 Julia, 194, 201 Knickerbock, 201 Kossuth, 204 Monte Cristo, 186 Occidental, 192 Ophir, 193, 200 Savage, 191, 194, 200, 201 Sierra Nevada, 197 Silver Queen, 246 Union, 185 Yellow Jacket, 192, 194, 196, 201, 207

UTAH: Bingham, 236 Copperfield, 238 Daly West, 237 Highland Boy, 238 Jordan, 236 Lark, 236 Little Bell, 237 Mercury, 238 Ophir, 236, 238 Park City, 236 Silver King, 237

Mines, Schools of, 114, 115, 153, 178, 219, 262

Minnesota, 131, 216, 127

Mississippi River, 22, 28, 44, 45, 52, 60, 61, 209, 210, 212, 222, 226, 233, 242

Missouri, 29, 93, 131, 211

Montana, 104, 109, 125, 130, 131, 140, 180, 189, 216, 222, 230, 238-248, 252, 258, 259

Mormons, 108, 185, 211, 212, 233-6

Morton, Charles, 14

Nebraska, 222

Nevada, 72, 93, 105, 111, 125, 130, 180, 183-207 *passim*, 208, 216, 225, 231, 246, 247

Nerve, Wilfred, 140-1

Newfoundland, 45

New Jersey, 14, 18, 76, 87, 229, 248, 252

New Mexico, 76, 109, 132, 175, 181, 230, 232, 248, 255, 257, 263, 268

New Orleans, 22, 29, 45, 60, 61, 233, 263

New York, 22, 37, 44, 95, 116, 120, 126, 127, 146, 148, 168, 173, 181, 225, 227, 237, 263, 268

Newspapers, 28, 30, 46, 54, 55, 58, 63, 64, 68, 69, 72, 74, 75, 76, 85, 88, 97, 108, 129, 131, 132, 141, 152, 158, 159, 161, 166, 197, 201, 203, 221, 223, 225, 236, 244, 254, 255, 256, 257, 259

Nicaragua, 52

Ohio, 14, 139, 231

Ohio River, 22, 44

Oregon, 30, 45, 72, 146, 208-218, 245, 250

Oregon Trail, The, 21, 211-14, 216, 222, 223

Panama, 22, 58, 59, 60, 61, 67, 75, 130, 229

Pearce, Bishop William, 146-8

Pearce, Richard B., 59, 153-4

Pennsylvania, 76, 77, 86, 107, 126, 139, 152, 228

Penrose, Charles 15

Penrose, Spencer, 181

Peru, 59, 86, 138

Philadelphia, 15, 32, 105, 115, 126, 147

Pittsburgh, 138

Placer mining, 51, 55, 62, 222

Place-Names:

CORNWALL: Ashton, 140 Barippa, 153 Blackwater, 109 Blisland, 32 Bodmin, 53, 201 Boskenwyn, 148 Breage, 91 Bugle, 263 Callington, 148, 182, 263 Calstock, 156 Camborne, 20, 21, 29, 30, 31, 37, 39, 44, 45, 50, 60, 72, 85, 93, 114, 153, 168, 176, 177, 194, 196, 197, 227, 232, 245, 263, 264, 265, 267 Camelford, 32, 34, 45 Carharrack, 150, 232 Carn Brea, 189 Carmeal Downs, 15 Chacewater, 93, 127, 148, 237 Charlestown, 14 Constantine, 40, 41, 43 Crowan, 200 Egloshayle, 14 Falmouth, 20, 30, 37 Fowey, 143 Golant, 143, 144 Grampound, 250, 252 Greenbottom, 58 Gweek, 32, 216 Gwennap, 49, 55, 149 Gwinear, 229 Gulval, 32 Gunnislake, 182, 204 Hayle, 55, 139, 143, 247, 263 Heamoor, 216, 230 Helford, 113, Helston, 15, 25, 32, 77, 102, 104, 132, 144, 180, 203 Illogan, 56, 139, 232 Kea, 49

Land's End, 29, 85, 217 Lanner, 27, 104, 105 Launceston, 14, 15 Lelant, 58 Lostwithiel, 53 Liskeard, 115, 216 Lizard, The, 10, 20, 35, 37, 39, 40, 43, 44, 144, 209, 215 Ludgvan, 32 Madron, 148 Marazion, 77 Mevagissey, 184 Mithian, 115 Morvah, 116 Mount Hawke, 225 Mullion, 21, 35, 36, 39, 43 Nancherrow, 156 Nancledra, 77, 93 Newlyn, 52, 144, 145, 201, 215, 219, 245 New Mill, 148 Newquay, 118 Paul, 235 Pengegon, 176 Pensilva, 115 Perranarworthal, 93 Perranporth, 174, 175, 237, 238 Perranzabuloe, 28, 61 Penryn, 14, 144 Penzance, 53, 55, 60, 102, 115, 116, 143, 155, 215, 216, 217, 237, 251, 255, 259, 263, 265 Phillack, 173 Ponsanooth, 58 Porthleven, 32, 36, 39, 144, 145 Porthtowan, 52, 59, 75, 77, 182 Redruth, 14, 15, 20, 55, 58, 75, 76, 85, 86, 104, 130, 150, 152, 161, 189, 223, 225, 227, 232, 246, 252, 259, 263, 265, 266, 267 Roche, 75, 149, 263 Roseworthy, 31 Rosmellen, 233 Ruan Major, 40 St. Agnes, 236 237, 264 St. Austell, 14, 26, 58, 149, 181, 216, 217, 230, 237, 260, 262, 263 St. Blazey, 161, 200, 263 St. Buryan, 79 St. Cleer, 79, 181 St. Columb Major, 15, 126 St. Day, 126, 149 St. Erth, 48, 51 St. Ewe, 28 St. Hilary, 58 St. Ive, 14 St. Ives, 46, 55, 127, 229, 263, 264 St. Just-in-Penwith, 17, 20, 58, 85, 102, 103, 115, 116, 132, 148, 154, 155, 174, 182, 191, 216, 219, 262, 263 St. Keverne, 38, 43, 209, 245 St. Mabyn, 14 St. Michael's Mount, 58 St. Neot, 115 St. Stephen, 14 Sancreed, 237 Seworgan, 216 Stithians, 180, 236 Stratton, 45 Townshend, 146, 147, 247 Tregony, 138 Trelan, 38, 209 Trethurgy, 263 Treswithian Downs, 29, 93, 98 Trewint, 14 Truro, 27, 58, 148, 153, 229, 237, 242, 263 Tuckingmill, 56, 115, 200 Tywardreath, 58, 87 Wadebridge, 251 Wenk, 113 Wendron, 233

ARIZONA: Ajo, 254 Apache San Carlos Reservation, 251, 255, 266, 268 Bisbee, 131, 168, 181, 237, 250-1, 260, 261, 262, 268 Cochise, 253 Florence, 259 Gila River, 253 255, 259, 268 Globe, 177, 237, 250, 251-2, 253, 255-269 Hayes Mountains, 259 Jerome, 168, 262, 263-4 Kofa Mountains, 253 Miami, 251, 261, 264-9 Mohawk, 253 Morenci, 263, 268 Pascoe Hill, 251, 268 Phoenix, 112,

248-9, 254 Pierce, 15, 253 Pinal Mountains, 251, 255 Prescott, 55, 248, 252, 254 Sacaton Mountains, 249 Saguaro Forest, The, 249 San Carlos, 256, 266 St. Just Hill, 15, 251 Superior, 254 Superstition Mountains, 249 Tempe, 256 Tombstone, 168, 253 Tubac, 254 Tucson, 232, 248, 255, 257 Wickenburg, 249, 254 Willcox, 259 Yuma, 248, 249, 251, 260
CALIFORNIA: Allison's Ranch, 55, 69, 221 Amador City, 70 American River, 51 Annaheim, 150 Auburn, 51 Barstow, 107 Bodie, 204 Boston Ravine, 69 Briceburg Canyon, 112 Burbank, 150 Calaveras River, 59, 103 Calico, 108-110, 130, 184 Cisco, 67 Columbia, 103 Cornish Mill, 67 Coulterville, 59, 113, Death Valley, 103, 110-112, 203 Deer Creek, 60, 67, 68 Downieville, 74, 78 Eureka Mills, 155 Feather River, 51, 78 Fresno, 87, 150 Gold Flat, 54, 56 Grass Valley, 25, 54-79, 84, 86, 87, 91, 102, 103, 106, 109, 185, 188, 190, 199, 204, 219, 221, 229, 232, 255, 265, 266 Inglewood, 150 Ione, 146 Jacksonville, 103 Jamestown, 107 Lake Tahoe, 183 Loomis, 51 Long Beach, 76 Los Angeles, 26, 30 72, 143, 150, 267 Los Gatos, 86, 93, Mariposa, 51, 62, 112, 113 Maryville, 69 Merced, 112 Merced River, 113 Mojave Desert, 103, 107-110, 130, 184 Mother Lode, The, 15, 22, 51, 52, 60, 61, 68, 76, 84, 87, 103, 112, 130, 146, 208, 236, 248, 258 Napa, 73 Nevada City, 54-79, 204, 255 New Almaden, 72, 79-103, 106, 107, 109, 128, 203, 258 New Helvetia, 50, 51 Oakland, 58, 67, 77, 78, 80 Panamint, 110, 203 Pasadena, 76 Placerville, 59, 61, 185 Randolph Flat, 68, 73 Redding, 250 Rough 'n Ready, 54, 60, 67 Sacramento, 50, 51, 58, 60, 62, 73, 76, 185, 186 Sacramento River, 22, 50 San Bernadino, 248 San Diego, 51, 248 San Francisco, 9, 15, 22, 25, 52, 53, 58, 60, 67, 72, 78, 80, 81, 86, 87, 93 (earthquake), 96, 98, 104, 105, 106, 111, 112, 185, 188, 191, 193, 213, 215, 246, 248, 259, San José, 79, 80, 81, 85, 86, 87, 93, 95, 98, 101, 102, 112, 146, Santa Clara, 51, 79-103, 204 Santa Cruz, 86 Sierra Buttes, 74, 78 Smartsville, 77 Sonoma, 50, 51, 52 Sonora, 60, 76, 103, 106, 107, 109 Soulsbyville, 103, 106, 109 Spanish Creek, 68 Springfield, 60 Stanislaus River, 59, 103 Sutter's Creek, 51, 75

Truckee River, 61 Tulane, 150 Tuolumne County, 103, 107 Vallejo, 50 Yosemite, 103, 113 Yuba River, 62
COLORADO: Aspen, 155-6 Black Hawk, 153, 156, 159, 163, 164, 168 Boulder, 153, 158, 181 Caribou, 164, 174, 176-7, 231 Cattle Creek, 155 Central City, 25, 148, 153-175 passim, 219, 251, 258 Clear Creek Canyon, 156, 161 Colorado River, 152 Colorado Springs, 176, 178 Cripple Creek, 180-2, 227, 253, 262 Denver, 153, 154, 156, 158, 161, 163, 168, 174, 175, 219 Fruita, 152 Georgetown, 153, 158, 177, 178 Gilpin County, 153-7, 160, 162, 163 Glenwood Springs, 155 Golden, 156, 158, 164, 217 Gregory Gulch, 25, 156, 158 Guston, 266 Idaho Springs, 156, 157, 158, 161-2, 163, 175, 182 Leadville, 74, 75, 155, 179-80, 182, 256, 257 Mount Wilson, 151 Nevadaville, 157, 158, 163, 164, 167, 174, 250 Ouray, 151, 266 Pike's Peak, 125, 153, 158 Pueblo, 176 Russel Gulch, 154, 156, 231 Silverton, 176, 266 Silver Plume, 178, 181, 182 Virginia Canyon, 156

DAKOTAS, THE: Belle Fourche, 149 Black Hills, 148, 149 Custer, 149 Deadwood, 148, 149, 259 Lead, 148, 149 Spearfish, 149 Terraville, 149 Terry, 217

IDAHO: Atlanta, 233 Banner, 233 Bellevue, 230-1 Boise, 68, 221, 222, 223, 224, 225, 226, 228, 230 Bunker Hill, 232 Burke, 219 Clearwater River, 219, 222 Coeur d'Alene River, 219, 222, 232 De Lamar, 228 Dewey, 228 Emmet, 230 Idaho City, 111, 222, 223 Owyhee Mountains, 223, 228 Pend d'Oreille River, 222 Quartzburg, 111 Rocky Bar, 223-8 Rupert, 230 Saw-Tooth Mountains, 223 Silver City, 72, 111, 228-9, 250 Snake River, 208, 212, 222, 223 Spokane River, 222 Sweetwater River, 212 Wallace, 219 Wardner, 219

MICHIGAN: Bessemer, 139 Bruce Crossing, 115 Calumet, 115, 117, 128, 130, 148, 246, 252 Copper Falls, 120 Detroit, 9, 29, 126, 127, 144, 230 Eagle Harbour, 120 Eagle River, 118, 120, 123, 126, 127, 132, 157 Garrison Dam, 131 Gogebic County, 139 Gratiot River, 124 Hancock, 56, 114, 115, 117, 126, 141, 262 Houghton, 20, 72, 114, 115, 126, 131, 149, 262 Iron Mountain, 141, 142, 143, 181

Iron River, 155, 141 Ironwood, 114, 115, 141, 143 Ishpeming, 139, 140, 145 Keweenaw Peninsula, 25, 114, 138 *passim*, 139, 146, 229 Lake Gogebic, 115 Lake Linden, 137, 138 Lake Michigan, 20, 21, 138 Lake Superior, 10, 19, 30, 115, 116, 123, 126, 129, 132, 138, 151, 162, 217, 259, 261 L'Anse, 116, 117 Marquette, 93, 125, 132, 138, 139, 141 Mass, 115 Negaunee, 138, 139, 140 Ontonogon River, 117 Red Jacket, 138 Silver Islet, 129-30 Teal Lake, 138 Trimountain, 116, 144 Wakefield, 140

MONTANA: Butte City, 111, 130, 131, 149, 168, 173, 230, 232, 237, 241, 242, 243, 247 Ennis, 245 Grass Range, 246

NEVADA: Battle Mountain, 93, 232 Carson City, 104, 106, 185, 207 Eureka, 77, 180, 195 Gold Hill, 72, 184, 185, 186, 190, 193, 195, 197, 201, 216 Humboldt County, 72 Lyon County, 195 Mount Davidson, 185-207 *passim* Ormsby County, 195 Pioche, 76 Reno, 105, 198, 207, 258, Rhyolite, 111, 112 Ruby Hill, 231 Silver City, 184, 192, 204 Six-Mile Canyon, 183, 185 Tonopah, 111, 246 Tuscarora, 55 Virginia City, 55, 58, 70, 72, 77, 87, 93, 104, 184-207, 227, 229, 242, 250 Walley's Hot Springs, 183 Washoe, 185, 197, 198 Zephyr Flat, 188

OREGON: Cascades, The, 208, 219 Clackamas River, 214 Columbia River, 150, 208, 209, 213, 216, 222 Coos Bay, 208 Crater Lake, 208 Dallas, The, 212, 213 Eugene, 216 Glacier Lake, 208 Humboldt Basin, 30 Jacksonville, 216 Manzanita, 11, 217 Milwaukee, 213, 214 Mount Hood, 208, 213 Oregon City, 209, 214, 217, 218 Oswego, 208 Portland, 130, 146, 213, 214, 215, 217 Roseburg, 216 Salem, 146, 215, 216 Silverton, 216 Willamette River, 21, 208, 209, 213, 216, 217

UTAH: Bingham, 229, 236, 237, 238, 263 Bonneville Desert, 238 Copperton, 238, Grantsville, 237 Green River, 212, 237 Lark, 236-8 Ophir, 236 Park City, 236, 237 Salt Lake City, 108, 217, 233, 234, 235, 236, 237, 238, 243 Sego, 237 Tooele, 232 Wasatch Mountains, 238

WISCONSIN: Beloit, 143, Caledonia, 38, 39, 43, 210 Cox Hollow, 32 Dodgeville, 31, 32, 45, 46, 47 Eagle, 34 Ferguson's Fort, 28 Frederick, 30 Galena, 22, 28, 44, 45, 46, 48, 60, 256 Hazel Green, 46, 47 Illyria, 36 Janesville, 32 Jenkinsville, 32 Kenosha, 36, 37 Kettle Moraine State Forest, 34 La Fayette County, 30 Linden, 31, 46, 47, 48 Little Prairie, 34 Madison, 36, 39 Melendy's Prairie, 46 Milwaukee, 34, 229 Mineral Point, 29, 31, 43, 46, 47, 48, 61, 63, 75 Palmyra, 32, 34, 46 Platteville, 46 Polk County, 30 Potosi, 32 Racine, 36, 37, 41, 48 Root River, 38, 210 Shullsburg, 29, 30, 32, 45, 46, 60 Whitewater, 32, 46, 48, 227 Yorkville, 36, 37, 38, 40, 41, 42, 43, 46, 60, 210

Plymouth, 34, 46, 53, 104, 145, 233, 251

Pony Express, The, 183, 238, 260

Population, size of Cornish mining, Cornwall, 16, 19
Arizona, 252, 253, 258, 263, 264
Calico, 108
Colorado, 162, 163, 179
Dakotas, the, 148
Grass Valley, 62, 63, 64, 67
Idaho, 221, 222, 223, 228, 229
Michigan, 119, 127, 131, 139, 141
Nevada, 186, 188, 191, 193
New Almaden, 82, 87, 88, 89, 92
Soulsbyville, 106
Tuolumne County, 107
Wisconsin, 46-7

Quebec, 22, 29, 36

Quicksilver, Methods of distillation, 83, 84

Railroads, 55, 67, 73, 93, 114, 116, 125, 130, 131, 132, 152, 156, 193, 195, 198, 236, 242, 257, 258, 259, 266

Randol, James B., 84 86, 87, 88, 89, 91, 92, 204

Religion, Cornish Ministers of, 41, 93, 115, 117, 123, 142, 143-8, 149-50, 154, 201, 215, 225, 245

Republicans, Cornish, 32, 75, 101, 131, 175, 230, 242, 243

Rickard, T. A., 59, 182

Robberies, 141, 190, 194, 249, 251

Rocky Mountains, 152-182, 266

Roman Catholics, 101, 222, 242, 243

Roosevelt, Theodore, 105

Rule, John (Real de Monte mine, Mexico), 59

Salivation, 96

Sangamon County (Illinois), 14

Schneider, James, 80

Sherman, 31

Ships, 36, 37, 41, 43-4, 45, 52, 60, 116, 126, 156, 174, 231, 235, 237, 263

Shootings, 72, 73, 76, 101, 102, 119, 157-8, 162, 165, 166, 176, 196-7, 199,